NEW TRENDS IN 3D PRINTING

Edited by **Igor V Shishkovsky**

New Trends in 3D Printing
http://dx.doi.org/10.5772/61398
Edited by Igor V Shishkovsky

Contributors

Daniel Guenther, Florian Moegele, Anne Isabelle Mertens, Jacqueline Lecomte-Beckers, Koichi Yano, Igor Shishkovsky, zengxi pan, Donghong Ding, Dominic Cuiuri, Huijun Li, Stephen Van Duin, Kaida Xiao, Julian Yates, Faraedon Mostafa, Ali Sohaib, Sophie Wuerger, Vadim Sufiiarov, Anatoliy Popovich, Heron Werner, Jorge Roberto Lopes Dos Santos, Leonardo Frajhof, Bruno Alvares De Azevedo, Luiz Lanziotti, Elyzabeth Avvad Portari, Sidnei Paciornik, Haimon Diniz Lopes Alves, Guangxue Chen, Hao Yin, Chen Chen, Liuxi He, Jiangping Yuan, Giovanni Biglino, Claudio Capelli, Andrew Taylor, Silvia Schievano, Pacurar Razvan, Pacurar Ancuta

Published by InTech

Janeza Trdine 9, 51000 Rijeka, Croatia

Publishing Process Manager Ana Simcic

Technical Editor SPi Global

Cover InTech Design team

First published July, 2016

Printed in Croatia

Additional hard copies can be obtained from orders@intechopen.com

New Trends in 3D Printing, Edited by Igor V Shishkovsky

p. cm.

Print ISBN 978-953-51-2479-5
Online ISBN 978-953-51-2480-1

Meet the editor

Igor Shishkovsky is Leading Researcher in Lebedev Physical Institute (LPI) of Russian Academy of Sciences (RAS), Samara branch, Russian Federation (RF). He holds a Doctor of Sciences degree in the chemical physics (including physics of combustion and explosion) from the Institute of Structural Macrokinetics and Material Science of RAS, Chernogolovka, and PhD in solid state physics from LPI of RAS, Moscow. He was teaching as Professor at Samara State Technical University and Moscow State Technological University and as Invited Professor at Ecole Nationale d'Ingenieurs de Saint Etienne, France.

Contents

Preface

A quarter century period of the 3D printing technology development affords ground for speaking about new realities or the formation of a new technological system of digital manufacture (DM), replacing the existing system based on the use of computers, Internet and nanotechnologies.

During this period, the 3D printing has undergone numerous significant changes on its way (Figure 1) resulting in an extra accuracy, enhanced mechanical features, broader scope and reduced costs of installations and of 3D parts and tools made by them. The 3D printing advantage is not only the arbitrariness of the shape but also the possibility of its instantaneous transfer to any point of the world that allows the industrial engineering on a world-wide scale. Cloud technologies and digital approaches have permitted a digital conceptualization for consumers, product creation and/or distribution. Successfulness of enterprises established on the 3D printing base is explained by their ability to recognize and estimate the market demand, which can be satisfied by creative supply of digital products.

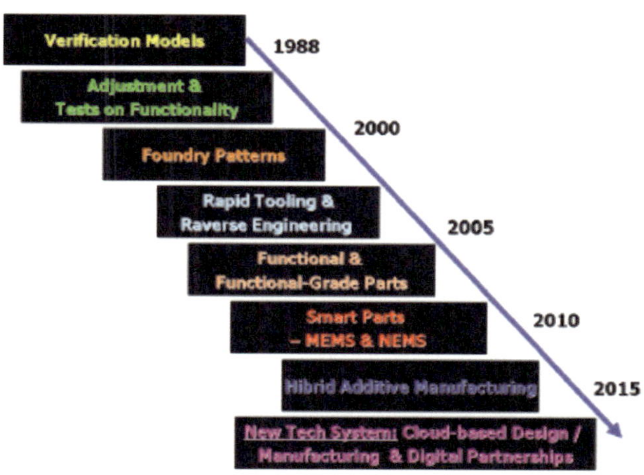

Figure 1. 3D printing progress from prototypes through functional tools and 3D parts to new technological system and digital partnership.

In our opinion, the up-to-date 3D printing is at the top of its own overrated expectations. In spite of the fact that many customers consider the 3D printing methods as "breaking through" and crucial for the new technological revolution (technological system), their influence in terms of the world-wide manufacture is so far insignificant. The development of scalable, high-speed methods of the material 3D printing aimed to increase the productivity and operating volume of the 3D printing machines requires new original decisions.

In recent years a close attention is paid to hybrid systems of the 3D printing, capable of creating in a unified technological process of finished structural 3D parts with built-in electronic, biological and /or chemical components, placed there by direct deposition or direct writing that allows along with the CAD-CAM of components, to make completely integrating electro-(/bio-) mechanic products (micro/nano-electro-mechanical systems — M/NEMS) as holistic or integral systems (Figure 2). Well known are the studies in such directions as optical MEMS (photon crystals, mirrors, optical switches, connectors, lens, PZT actuators etc), bio-MEMS (implants, scaffolds etc), microfluidic MEMS (drug delivery systems, lab-on-chip, pumps, catalyst membranes etc), power MEMS (fuel cells and batteries, energy harvesting, magnetic devices etc), and radiofrequency MEMS (antennas, resonators & oscillators, actuators, phase shifter, band-pass filters etc). However, the tasks of the MEMS fabrication with feedback control, adaptive management and possibilities of prediction and response for the 3D printing processes still demand their solutions.

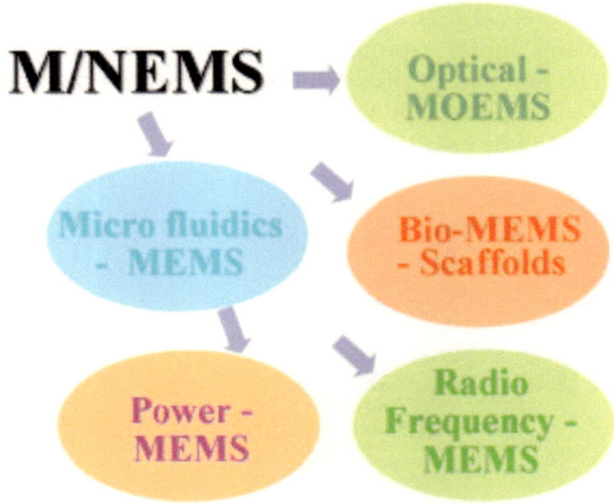

Figure 2. Scope of micro/nano-electro mechanical systems.

Materials are also an essential and integral part of the 3D printing technologies. The key problem for the materials design, manufacturing and processing is the improvement of their quality, expansion of the suitable materials spectrum (due to mixing/ alloying/ composite's modeling), increase of the process stability, repeatability and reliability for multi-material systems, with the retention of a low cost of materials, the process of their fabrication and pre- and/or post- processing.

For a great number of materials and the 3D printing processes large-scale studies are required, as well as the determination of correlations "process–structure–feature". Of not less interest are the problems of using the unique possibilities of the 3D printing for creation of the unique structures not existing in nature or reproducing its best qualities (e.g., parts with negative coefficient of thermal expansion or electromagnetic wave transmission), metamaterials, biomimetic composite constructions and surfaces.

It is necessary to study the 3D printing applicability for manufacturing of the materials with multilevel hierarchical functionality on nano-, micro- and meso-scales and also the development of instruments for fabrication of 3D printing structures via atom-by-atom approach, and design of additive nanofabricators.

Some of the above mentioned problems and issues are considered in this book which consists three parts. The first part offers advanced approaches for the 3D printing methods. The second part is completely dedicated to medical applications of the 3D printing where the main advantage of additive technologies is realized, i.e. the ability to use individual data of each patient, herewith solving the problems of the products personalization. It's probable that among readers are those occupied with studies of layerwise approaches to fabrication of the human organs, and we hope that some of the authors' ideas described in this part of the monograph could be useful for them. Finally, the third part of the monograph presents some new approaches for the powdered methods of the 3D printing that can also find applications for medical, aerospace and/or automotive industries, always being the key branches facilitating the innovation development in the additive manufacture.

Prof. Igor V. Shishkovsky
Laboratory of Laser Technologies,
Samara branch of P.N. Lebedev Physical Institute,
Russian Academy of Sciences
Samara, Russian Federation

Section 1

Advanced 3D Printing

Chapter 1

Advanced Design for Additive Manufacturing: 3D Slicing and 2D Path Planning

Donghong Ding, Zengxi Pan, Dominic Cuiuri,
Huijun Li and Stephen van Duin

Additional information is available at the end of the chapter

http://dx.doi.org/10.5772/63042

Abstract

Commercial 3D printers have been increasingly implemented in a variety of fields due to their quick production, simplicity of use, and cheap manufacturing. Software installed in these machines allows automatic production of components from computer-aided design (CAD) models with minimal human intervention. However, there are fewer options provided, with a limited range of materials, limited path patterns, and layer thicknesses. For fabricating metal functional parts, such as laser-based, electron beam-based, and arc-welding-based additive manufacturing (AM) machines, usually more careful process design requires in order to obtain components with the desired mechanical and material properties. Therefore, advanced design for additive manufacturing, particularly slicing and path planning, is necessary for AM experts. This chapter introduces recent achievements in slicing and path planning for AM process.

Keywords: Additive Manufacturing, GMAW, GTAW, Path Planning, 3D Slicing, Welding, Medial Axis Transform, STL

1. 3D CAD slicing

AM requires an input computer-aided design (CAD) model of the part which may be designed in a CAD system, or obtained from reverse engineering such as 3D scanners. Once the CAD model is completed, it is converted to the standard STL format, which is most commonly used to represent 3D CAD models in additive manufacturing. A section of an STL file and its 3D model are shown in **Figure 1**. In an ASCII STL file, the CAD model is represented using triangular facets, which is described by the x-, y-, and z-coordinates of the three vertices and a unit vector to indicate the normal direction that point outside of the facet [1]. Redundant information for indicating model name, facet normal, and vertex is also included. In the following sections, unidirection slicing and multidirection slicing of STL models are described.

```
solid CREO STL
 facet normal 0.000000e+00 0.000000e+00 -1.000000e+00
  outer loop
   vertex 5.000000e+01 -5.000000e+01 0.000000e+00
   vertex -5.000000e+01 5.000000e+01 0.000000e+00
   vertex 5.000000e+01 5.000000e+01 0.000000e+00
  endloop
 endfacet
 facet normal 0.000000e+00 1.000000e+00 0.000000e+00
  outer loop
   vertex 5.000000e+01 5.000000e+01 6.000000e+00
   vertex 5.000000e+01 5.000000e+01 0.000000e+00
   vertex -5.000000e+01 5.000000e+01 0.000000e+00
  endloop
 endfacet
......
......
endsolid CREO STL
```

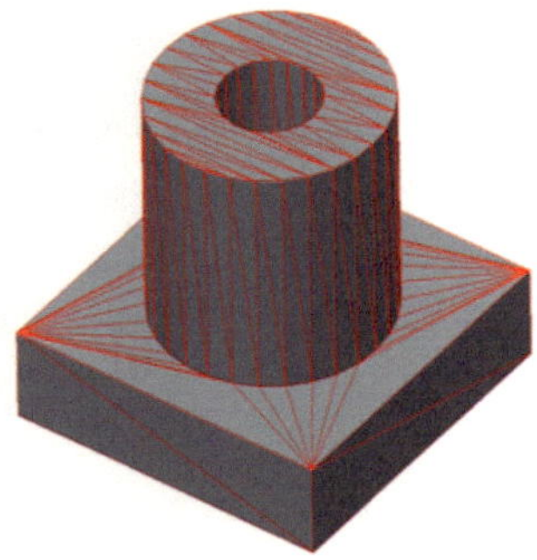

Figure 1. Example of an STL file format and its model.

2. Unidirection slicing

Unidirection slicing algorithm slices the STL model into a variety of 2.5D layers parallel to the build direction. **Figure 2a** shows an STL model and various cutting planes for unidirection slicing; and **Figure 2b** shows the resulting slicing layers which are polygons.

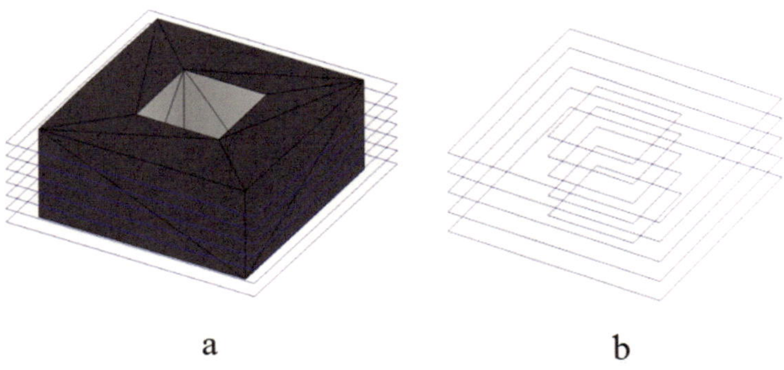

Figure 2. (a) STL model with slicing layers in unidirection; (b) Resulting layer boundary polygons.

Uniform slicing method slices CAD model with a constant thickness. The accuracy of the additively manufactured parts could be controlled by altering the layer thickness. The smaller the layer thickness, the higher the accuracy will be obtained. Also, the deposition rate or productivity is highly relevant to the sliced layer thickness. Adaptive slicing approach slices a model with a variable thickness. Based on the surface geometry of the model, this approach adjusts automatically the layer thickness to improve the accuracy or to improve the build time.

With the increasing size of the STL file, a major challenge of slicing algorithms is the computing efficiency. An efficient and fault-tolerant slicing algorithm was proposed by Choi et al. [2]. The detailed slicing algorithm is presented as follows.

Figure 3 shows the flowchart of the tolerant slicing algorithm [2]. Two stages are included, namely, preprocessing stage regarding the optimum orientation of the part, and slicing stage generating slices from STL model. The orientation of the STL model is usually user defined or obtained with respect to the optimal build time, the surface quality, support structures required, or other criteria. Algorithms regarding the orientation of CAD models are beyond the scope of this chapter.

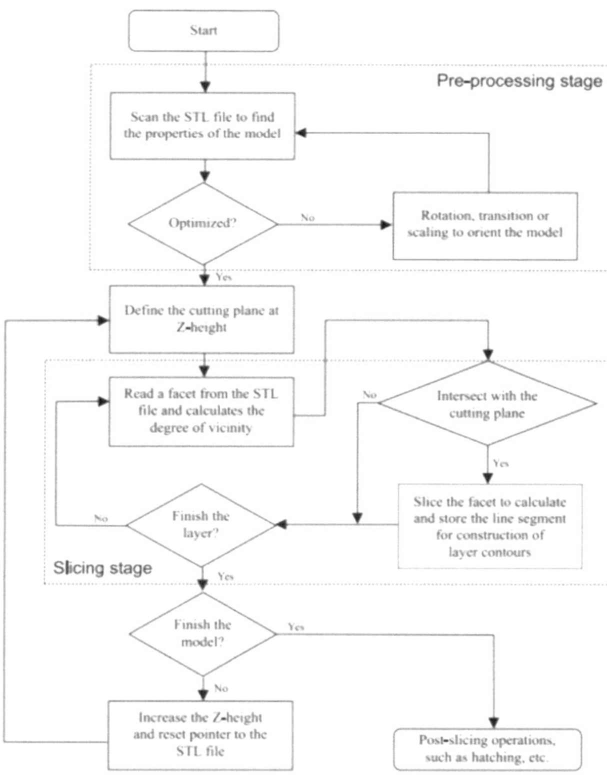

Figure 3. Flowchart of the tolerance slicing algorithm [2].

For each layer construction in slicing, the method scans the STL file to extract one facet at a time sequentially and compares the z-coordinates of its three vertices to the z-height of the slicing (cutting) plane. Through setting a small value bound tolerance, the possibilities of intersection of a facet with a cutting plane can be divided into the following four classes, as shown in **Figure 4** [2]. Four classes could be found as following.

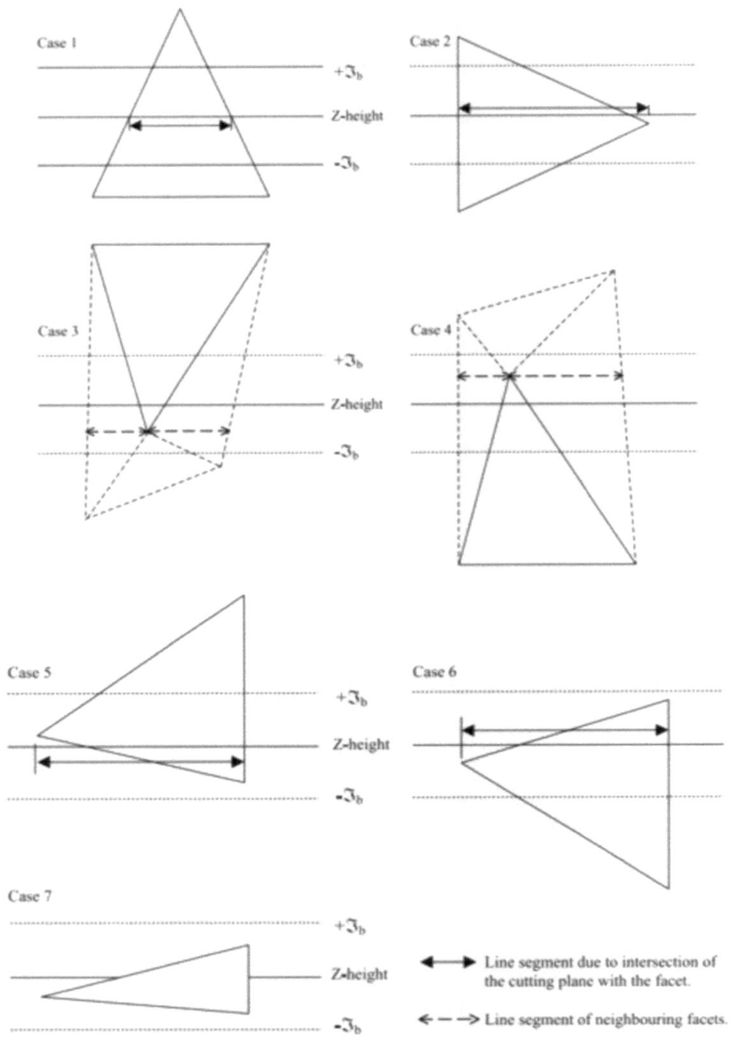

Figure 4. Possible cases of facet-plane slicing [2].

Class 1 is shown in case 1 in **Figure 4**, where a facet intersects with the cutting plane without one vertex lies on the plane. Two intersecting points are calculated by the intersection of the plane and the triangular facet in this case. A line segment is generated by connecting the intersection points and stored for constructing the slice layer contour [2].

Class 2 is shown in case 2 in **Figure 4**, where one vertex lies in the cutting plane while the two remaining vertices lie in different sides of the cutting plane. Case 3 and case 4 also show that one vertex lies in the cutting plane while other vertices lie in the same side of the cutting plane. In this case, there will be no intersection line segments calculated [2].

Class 3 is shown in case 5 and case 6 in **Figure 4**, where two vertices of the facet lie in the cutting plane. The remaining vertex can be either above or below the cutting plane. This is a concurrent case in a valid STL model that results in one duplicated line segment and either one should be ignored [2].

Class 4 is shown in case 7 in **Figure 4**, where all three vertices lie in the cutting plane. In this case, the whole triangular facet lies in the cutting plane and will not contribute any line segments to the contour of the slice layer. Therefore, this case can be directly ignored without any ambiguity [2].

After all the possible intersection line segments have been calculated and stored, the contour of the slice layer is constructed through head-to-tail connecting of the line segments.

The algorithm has been implemented in MATLAB program. The slicing algorithm was able to slice complex STL models with good performance. Three case studies are provided as shown in **Figure 5–7**. The number of facets of three models is provided in **Table 1**. The algorithm is verified to be efficient with the computing time less than half second for all cases.

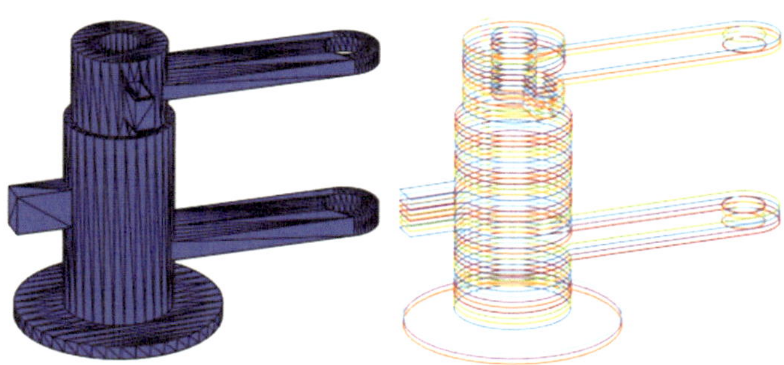

Figure 5. Case I. (a) An engineering STL model; (b) Resulting slices.

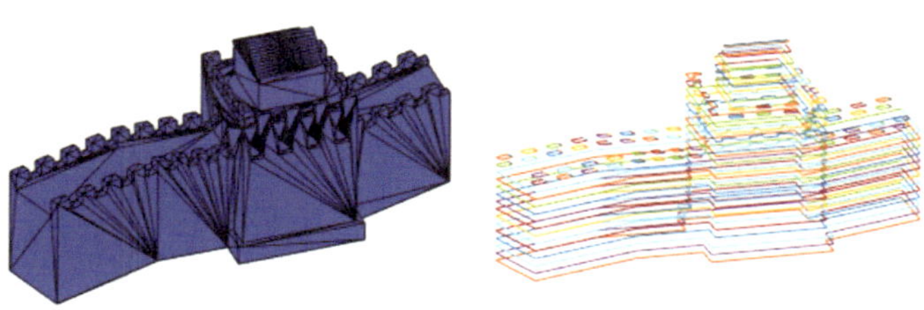

(a) (b)

Figure 6. Case II. (a) A Great Wall STL model; (b) Resulting slices.

(a)

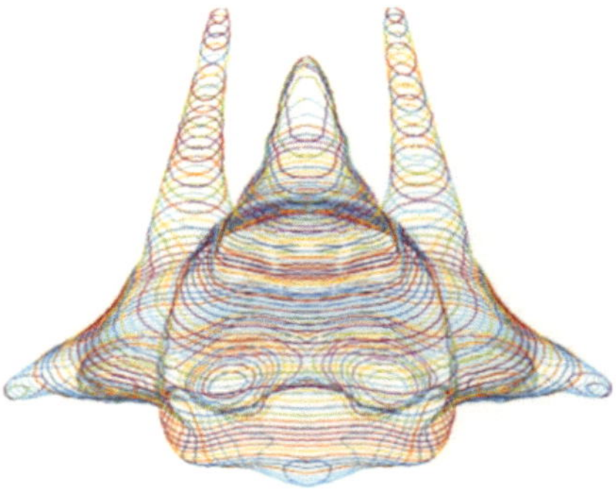

(b)

Figure 7. Case III. (a) A Triceratops STL model; (b) Resulting slices.

No.	Facets	Computing time
Case I	1076	12 ms
Case II	1584	34 ms
Case III	172,122	352 ms

Table 1. The computing time of the program.

3. Multidirection slicing

Most of the current AM processes involve slicing a 3D CAD model into a set of 2.5D layers with a constant or adaptive thickness along the build-up direction as mentioned in unidirection slicing. However, to fabricate parts with complex shapes the unidirectional slicing strategies are generally limited by the need for support structures to deposit overhangs. **Figure 8a** shows a component and its usual build direction *B*. It is clear that to fabricate the component in direction *B*, support structures (refer **Figure 8b**) are required due to overhangs. The deposition of supports results in the wastage of materials and the removal of these supports requires costly post-processing.

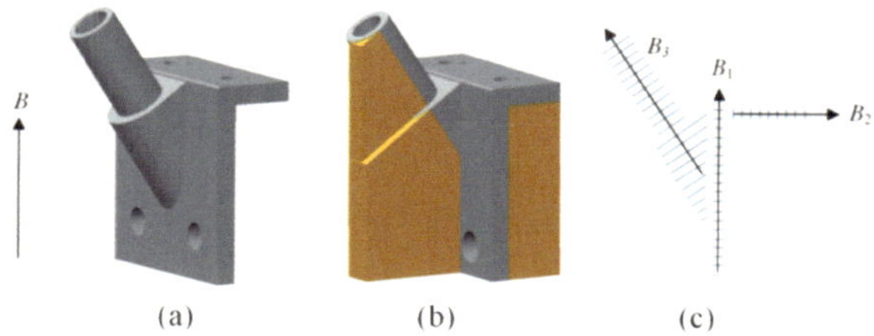

Figure 8. (a) The component and its usual build direction B; (b) support required (parts colored in orange); (c) multi-direction slicing and various build directions B_1, B_2, and B_3 [3].

Some of the additive manufacturing techniques are able to deposit materials along multiple directions. The application of multidirection deposition could eliminate or significantly decrease the usage of supports for complex components. As shown in **Figure 8c**, the component can be fabricated in multiple directions, for example, B_1, B_2, B_3, without support structures. Such a multidirection deposition system furthers the capability of layered manufacturing by reducing the need for supports.

A key challenge in multidirection AM is to develop robust algorithms capable of automatically slicing any 3D model into a set of layers which satisfy support-less and collision-free layered deposition. The following sections review some existing multidirection slicing strategies.

Silhouette edges projection [4]: This strategy firstly identifies the unbuildable surface features of a model by projecting silhouette edges along the user defined original build direction. Then the part is decomposed into buildable and unbuildable sub-volumes using the silhouette-edge-based method. For the unbuildable sub-volume, a new suitable build direction is determined using the Gauss and Visibility maps. With the new build direction, the unbuildable sub-volume is further decomposed through repeating the same projection procedures. This projection method is recursively used to decompose the sub-volume as shown in **Figure 9**. The framework for the multidirection slicing and some essential problems have been addressed and discussed by the authors. However, the implementation of the strategy could be complicated and computationally expensive for complex components with inner cavities.

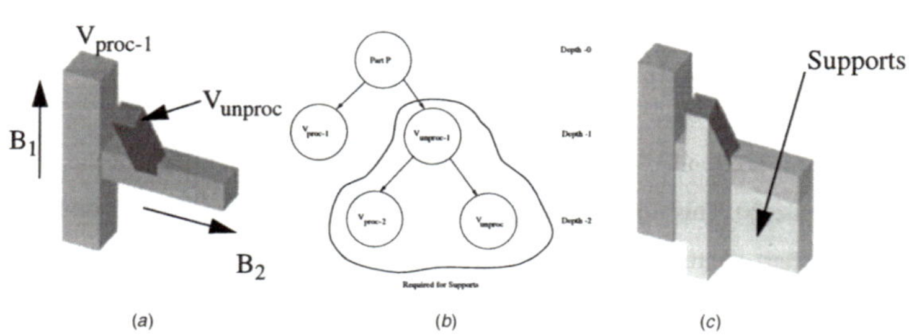

Figure 9. (a) Model; (b) Decomposition tree; (c) Supports [4].

Transition wall [5]: The key idea of this strategy is to identify the overhang layers by computing the difference between the current layer and the previous layer. Then, as shown in **Figure 10**, to build an overhang structure, the machine is turned 90° to start depositing a transition, namely thin wall. After the deposition of the first few layers, the wall is finished and the subsequent overhang structures can be deposited in the vertical direction again. Although this strategy is simple, it is only suitable for a subset of part geometries. In some cases, the deposition of the transition wall is difficult or impossible to implement due to deposition nozzle collision, such as the part shown in **Figure 8a**.

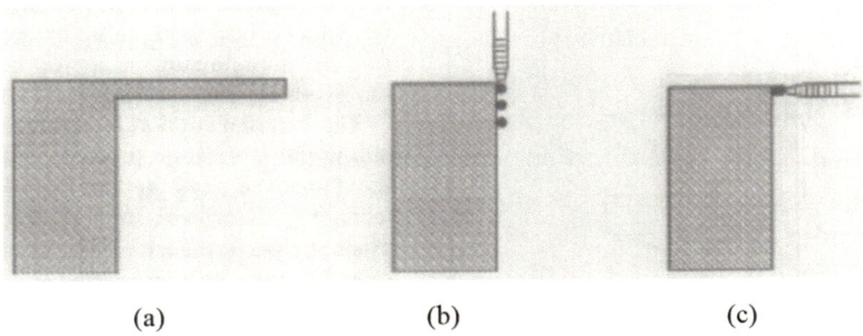

Figure 10. Illustration of building the transition wall through rotating 90 degree. (a) Overhang; (b) Vertical deposition; (c) Horizontal deposition [5].

Centroid axis extraction [6]: The first step in this strategy is to extract the centroid axis of the model as shown in **Figure 11**, which provides a global perspective on the geometry, allowing the slicing procedure to be conducted on an optimal sequence. Through analyzing the topological information from the centroid axis, the splitting surface is identified and the subsequent decomposition operation is conducted. For each subcomponent obtained from decomposition, multiaxis slicing is performed and the collision-free slicing sequence is finally generated. The centroid axis extraction method decomposes the component by detecting the change of centroid of presliced layers, making the geometry analysis process easier. However, in some cases it will be difficult to decompose components efficiently as required since the centroid axis does not always indicate the change of the geometry.

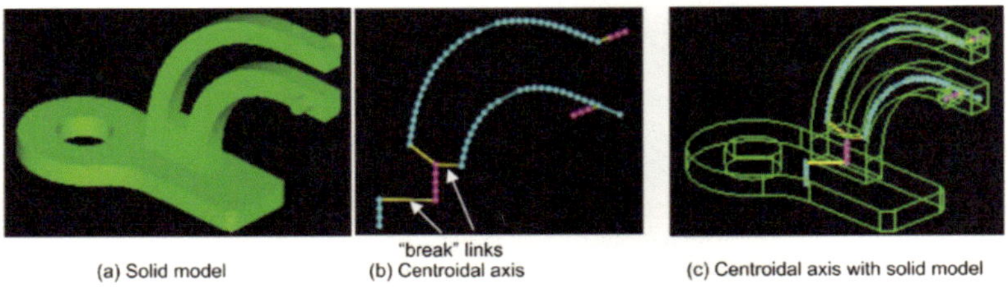

Figure 11. Example of centroidal axis extraction [6].

Other multidirection slicing methods have been proposed that are either adaptations or combinations of a few techniques from the above strategies, such as adaptive slicing algorithm [7], offset slicing [8], skeleton method [9], and modular boundary decomposition [10]. However, each method is only suitable for a subset of part geometries. In addition, these methods are not efficient for processing parts with holes and depression features. As shown in **Figure 12**, the part is decomposed into buildable volume and unbuildable volume due to the direction of hole, H, is vertical to the build direction, B, of its associated volume. The part is firstly decomposed into buildable and unbuildable sub-volume [4]. The unbuildable sub-volume could be further built by offset slicing strategy [8] with the direction as shown in **Figure 12b**. However, unbuildable volume also exists due to the holes. It is clear that holes would hamper implementing multidirection slicing algorithm.

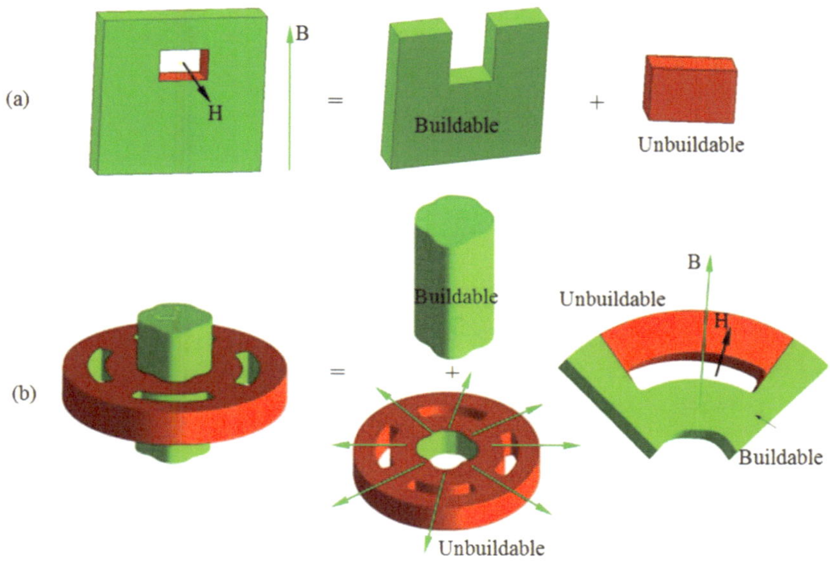

Figure 12. Illustration of the impact of holes on multidirection slicing strategy [3].

Decomposition–regrouping method [3]: Differing from these existing slicing approaches, which are mainly focused on finding an optimal volume decomposition strategy, a recent study proposed simplification of holes and a decomposition–regrouping method. A model simplification step is introduced before CAD decomposition to significantly enhance the capability of the proposed multidirection strategy. The CAD model is then decomposed into sub-volumes using a simple curvature-based volume decomposition method and consequently a depth tree structure based on topology information is introduced to merge them into ordered groups for slicing. The proposed strategy is proven to be simple and efficient on various tests parts. As an example shown in **Figure 13**, the part is decomposed and regrouped, then sliced in multiple directions with holes been filled. Since there is no robust multidirection slicing algorithm validated for any complex geometry, the proposed multidirection slicing algorithm is not versatile also, while it is particularly useful for components with large number of holes.

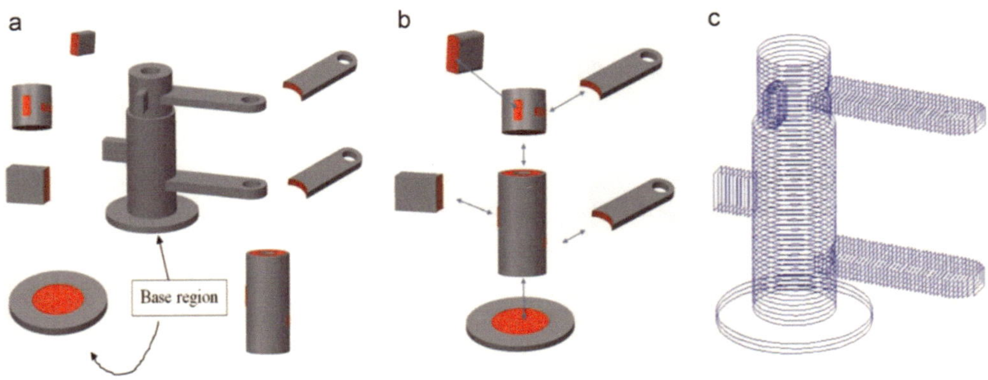

Figure 13. Example of decomposition–regrouping method for multidirection slicing. (a) Volume decomposition; (b) Sub-volume regrouping; (c) Slicing in multiple directions [3].

4. 2D path planning

Another important step in AM is the development of an elaborate path planning strategy. Path planning for powder-based AM processes that have fine, statistically distributed particles is somewhat independent of geometric complexity. However, path planning for AM processes that have coarse and large-sized deposits is influenced by geometric complexity. Also, the property of the deposited shape will be influenced by the deposition path trajectory. In the following sections, methods are described to generate different types of deposition paths.

5. Raster path

The raster scanning path technique, as shown in **Figure 14a**, is based on planar ray casting along one direction. In this strategy, 2D regions are filled by a set of scan lines with finite width [11]. It is commonly employed in commercial AM systems due to its simple implementation and suitability for almost any arbitrary boundary.

6. Zigzag path

Derived from the raster strategy, zigzag tool-path generation is the most popular method used in commercial AM systems. While it fills geometries line by line along one direction like the raster approach, the zigzag approach combines the separate parallel lines into a single continuous pass (see **Figure 14b**) which significantly reduces the number of tool-path passes [12, 13]. This method significantly improves the productivity of the AM process by reducing the required transition motions of the machine. However, the outline accuracy of the part for both raster and zigzag approaches is poor due to the discretization errors on any edge that is not parallel to the tool motion direction.

7. Contour path

Contour path generation as shown in **Figure 14c**, which is another typical method, can address the above geometrical quality issue effectively by following the geometrical trend of the boundary contours [14, 15]. Various contour map patterns were investigated by Li et al. [16] to develop optimal tool-path patterns for sculptured parts with a single island and no seriously non-convex shape.

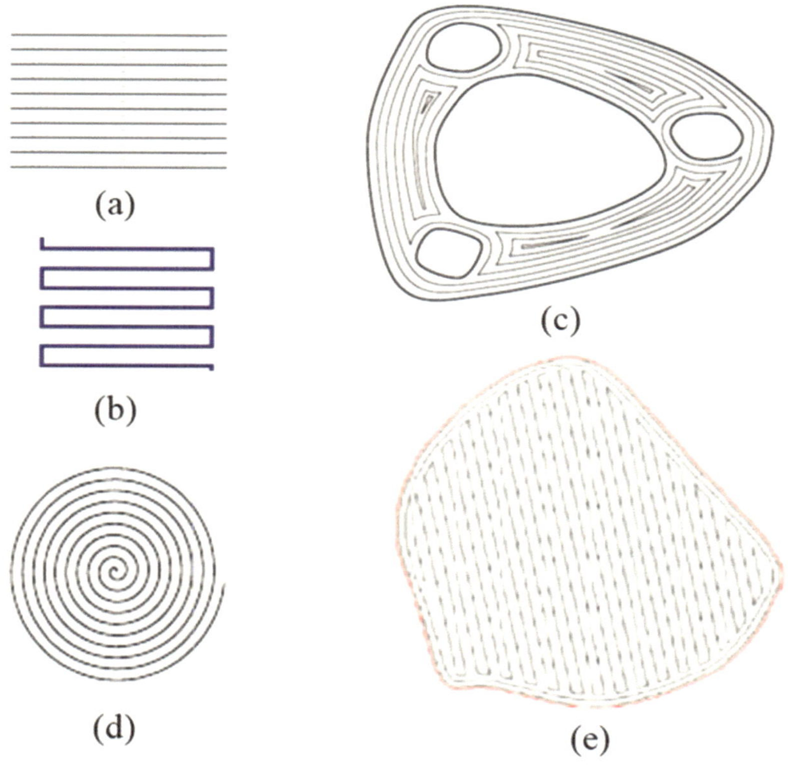

Figure 14. Raster path pattern. (a) Raster path; (b) Zigzag path; (c) Contour path [15]; (d) Spiral path [17]; (e) Hybrid path [18].

8. Spiral path

The spiral tool-path generation, as shown in **Figure 14d**, has been widely applied in numerically controlled (NC) machining, especially for 2D pocket milling and uniform pocket cutting [17, 19]. This method can also be used to solve the problems of zigzag tool paths in AM process, but is only suitable for certain special geometrical models.

9. Hybrid path

The hybrid path planning strategy is promising as it shares some merits of various approaches. A combination of contour and zigzag pattern is commonly developed to meet both the geometrical accuracy and build efficiency requirements. Zhang et al. [20] applied a new image algorithm for welding-based AM. The planned approach includes one inner zigzag path which is faster and more universal, and one outline vector path which is very helpful for maintaining the surface accuracy and quality. Jin et al. [18] proposed a mixed tool-path algorithm to generate contour and zigzag tool path for AM of biomedical models, as shown in **Figure 14e**. The zigzag tool path is employed to fill the interior area of the part to improve the efficiency, while the contour tool path is used to fabricate the area along the boundary of the contours to improve the geometrical quality of the model.

10. Continuous path

Continuous path planning can be considered as another tool-path generation method. Hilbert filling curve applied by Bertoldi et al. [21] is a continuous path, which can cover a region of space without intersecting itself as shown in **Figure 15a**. It has been found to be particularly useful in reducing shrinkage during AM fabrication processes. However, the large number of path direction turning motions that are produced by using this strategy are not suitable for AM. Wasser et al. [22] introduced a fractal-like build style using a simulated annealing algorithm. This method is able to generate filling patterns that allow the continuous deposition of a single path to fill arbitrarily shaped areas, as shown in **Figure 15b**. In this method, the area to be deposited is firstly decomposed into nodes, with the number of the nodes determined by the accuracy requirement of the AM process. However, when the area to be filled is large and the accuracy requirement is high, the processing time required would be unacceptably long. Moreover, highly convoluted paths may result in accumulation of heat in certain regions, therefore inducing excessive distortion of the part. Frequent alternations of tool-path travel directions are also not preferred in AM.

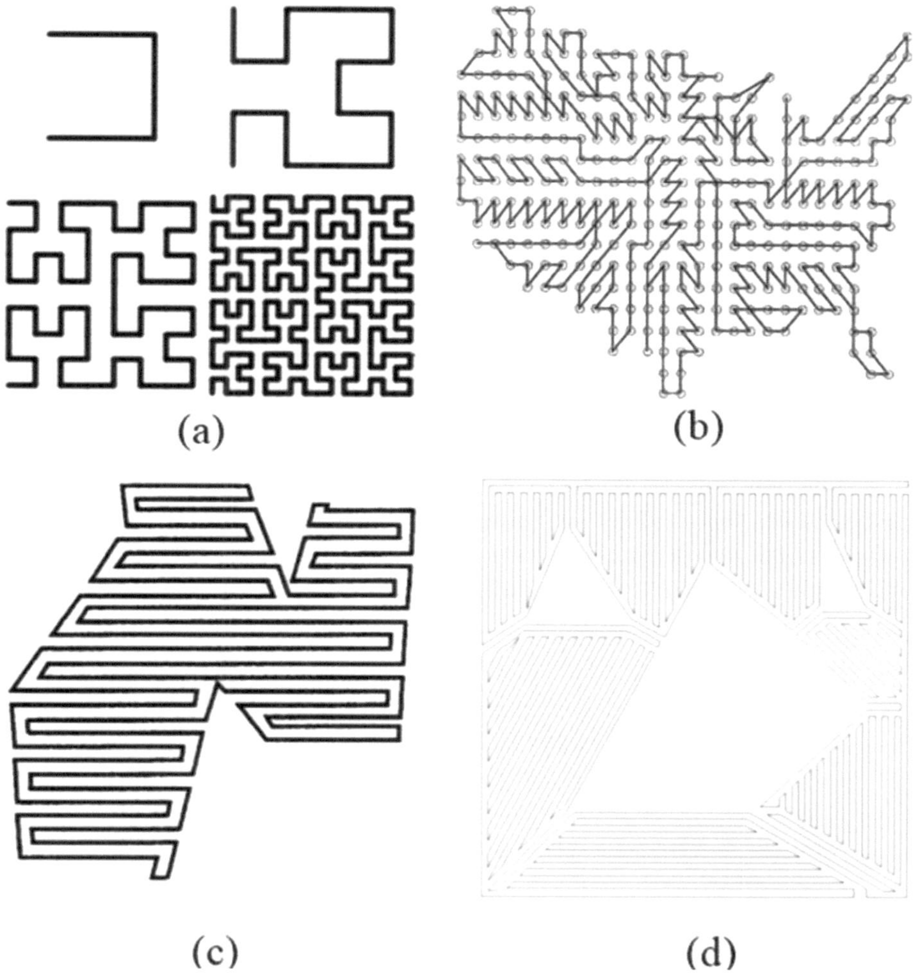

Figure 15. Continuous path pattern. (a) Hilbert filling curve [21]; (b) Fractal-like build style [22]; (c) Hybrid continuous path (zigzag dand continuous) [23]; (d) Hybrid continuous path (zigzag and contour an continuous) [24].

11. Hybrid and continuous path

Dwivedi et al. [23] developed a continuous path planning approach which combines the merits of zigzag and continuous path patterns, as shown in **Figure 15c**. In this method, the 2D geometry is firstly decomposed into a set of monotone polygons. For each monotone polygon, a closed zigzag curve is then generated. Finally, a set of closed zigzag curves are combined together into an integrated continuous torch path.

The continuous path planning approach significantly reduces the number of welding passes. However, purely employing zigzag-based curves would have a surface accuracy issue, as

Zhang et al. [20] point out the importance of filling the outline of the image with vector motions in wire and arc additive manufacturing (WAAM). Recently, we proposed another continuous path pattern which leverages the advantages of zigzag, contour, and continuous path patterns, as shown in **Figure 15d** [24]. It is indicated that this continuous path pattern is suitable for WAAM of solid structures.

12. Medial axis transformation (MAT) path

Kao [25] has proposed an alternative methodology of using the medial axis transformation (MAT) of the geometry to generate the offset curves by starting at the inside and working toward the outside, instead of starting from the boundary and filling toward the inside. This approach is able to compute paths which can entirely fill the interior region of geometry as the paths are generated from interior to the boundary. This strategy avoids producing gaps by depositing excess material outside the boundary, as illustrated in **Figure 16**. The extra material can subsequently be removed by post-processing. Therefore, the traditional contour path patterns from outside to inside is natural for machining whereas MAT path starting from inside and working toward the outside is suitable for AM of void-free components.

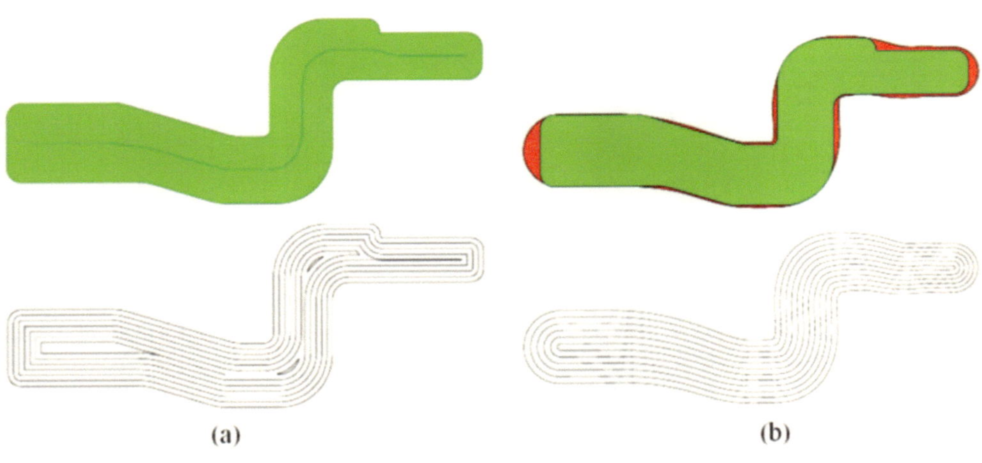

Figure 16. Illustration of path generated from MAT. (a) up: original geometry (green region) and the MAT of the geometry (black line); down: contour path patterns with gaps in middle are clearly seen; (b) up: red region is deposition of excessive materials; down: the MAT path patterns without gaps [25].

However, the previous authors limit their discussion to geometries with simple MAT paths. Recently, Ding et al. [26, 27] developed a methodology of generating MAT-based paths for an arbitrary geometry, either thin-walled or solid structures.

The following are the main steps for generating MAT-based paths:

Compute the medial axis: The cross section of a sliced layer with a thin-walled structure is shown in **Figure 17a**. The medial axis or skeleton of the geometry is computed and represented using red lines. The computed skeleton is the crucial information, which describes the shape of the geometry.

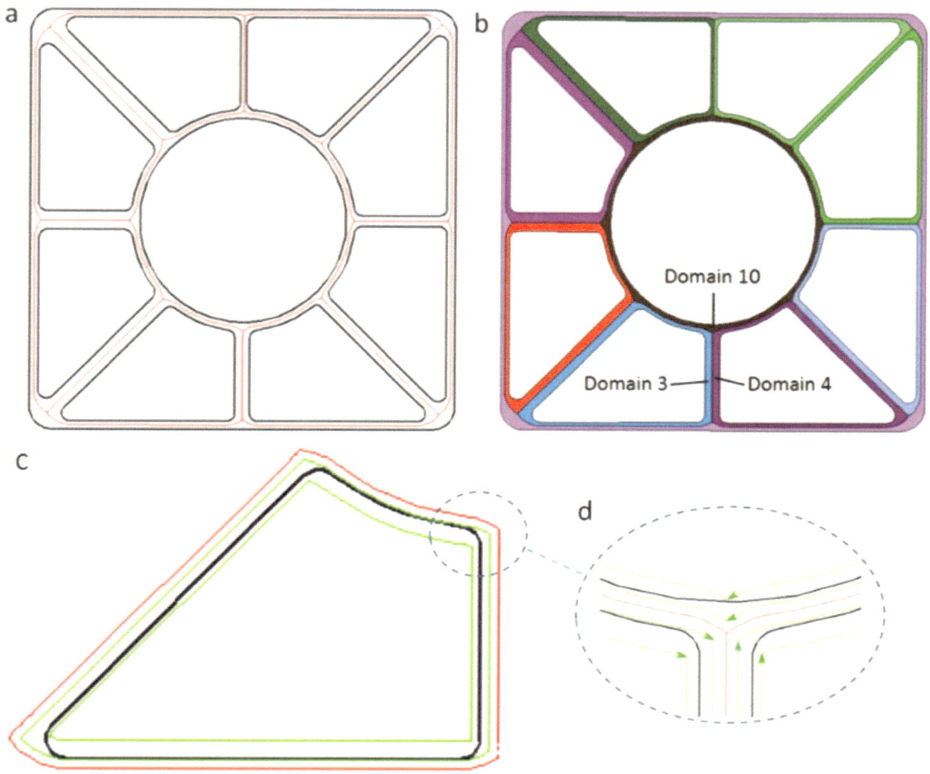

Figure 17. Illustration of MAT-based path planning. (a) The medial axis computing (red lines). (b) Domain decomposition (each domain is described in one different color. (c) Path generation for domain 3. (d) MAT-based paths at the crossing area [27].

Decompose the geometry: Using the computed medial axis, the geometry is decomposed into several domains. As the geometry with N holes will be decomposed into N+1 domains, this geometry is decomposed into 10 domains as shown in **Figure 17b**. As displayed in different colors, each domain is bounded by a portion of medial axis (red lines in **Figure 17a**) and a boundary loop (black line loop in **Figure 17a**).

Generate path for the domain: Deposition paths for each domain are generated by offsetting the medial axis loop (red line loop in **Figure 17c** for domain 3) toward the corresponding boundary loop (black line loop in **Figure 17c**) with an appropriate step-over distance. The offsetting is repeated and terminates when the domain is fully covered. Green line loops in **Figure 17c**, represent the generated deposition paths.

Complete the deposition paths: A complete set of MAT-based deposition paths is obtained by repeating step 3) for all the decomposed domains. The generated paths are a set of closed-loop lines without start/stop sequences, which is preferred for the arc welding system.

From the above description, the MAT path planning algorithm for an arc welding process is able to be automated for any complex geometry; just as the existing commercially available raster and contour path planning strategies have been automatically applied to powder-based AM. MAT path is particularly preferred for void-free AM. Example of MAT path generation for a solid structure with holes is shown in **Figure 18**.

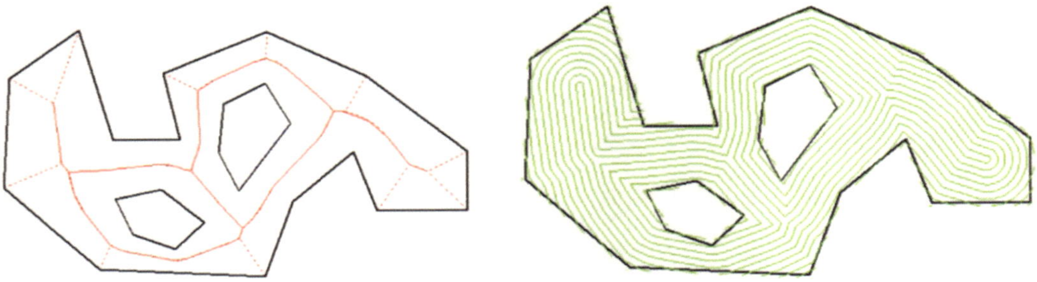

(a) (b)

Figure 18. Example of a solid structure with holes. (a) Geometry is represented by black lines, MAT represented by dotted red lines, and red solid lines stand for branches. (b) Generated trimmed path [26].

13. Adaptive MAT path

Traditional contour path always generates gaps or voids as shown in **Figure 19a** and **b**. To avoid internal voids, the MAT path was introduced and its extension for complex geometries was developed. MAT paths are generated by offsetting the medial axis of the geometry from the center toward the boundary. **Figure 19c** shows an example of MAT path with the deposition sequence indicated by numbers. Although void-free deposition is obtained using MAT paths, this is achieved at the cost of creating discontinuity of the path (such as path 3, 4, and 5 in **Figure 19c**) and extra deposition at the boundary as described in **Figure 19d**. Post-process machining must be used to remove the extra materials and improve the accuracy at the cost of material and energy wastage.

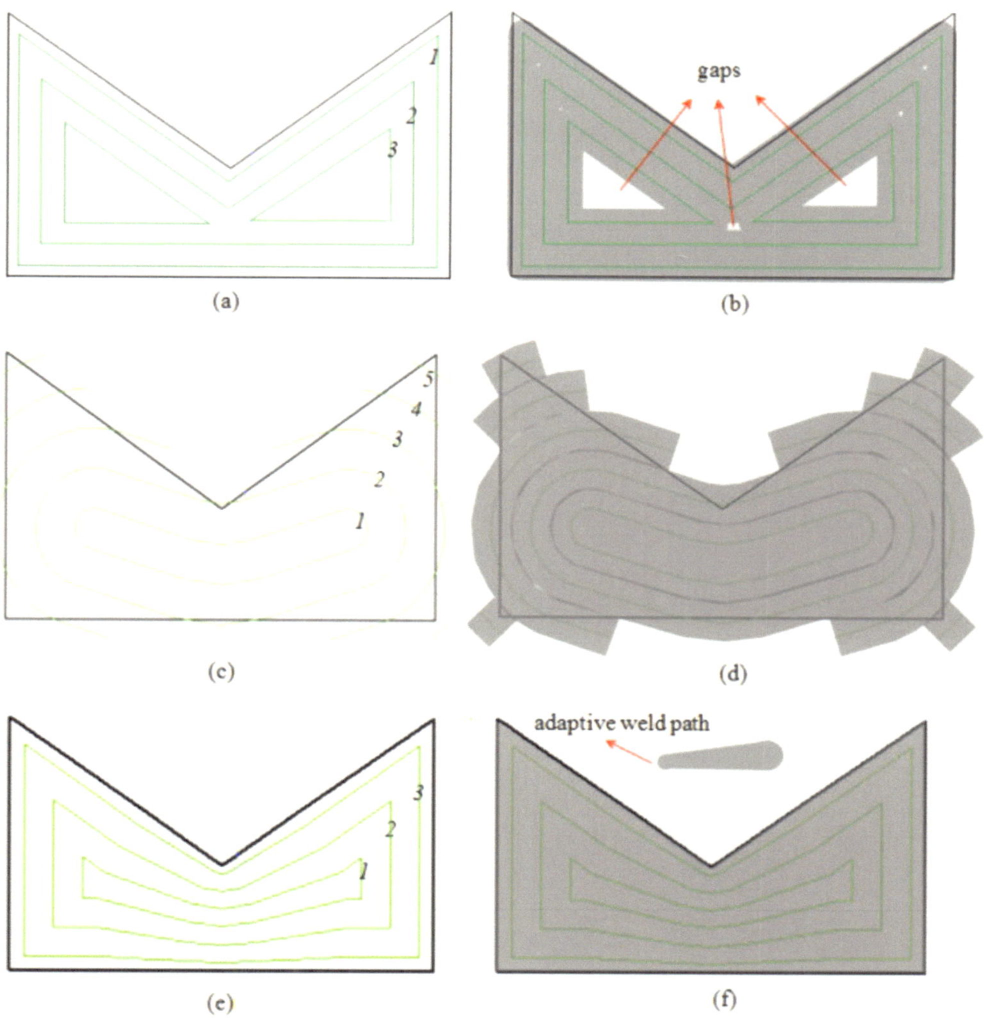

Figure 19. Illustrations of different deposition paths. Black lines represent the boundary of the geometry; green lines represent the deposition paths with the numbers representing the order of the deposition paths; grey regions are deposited area by the relevant paths. (a) Contour path patterns; (b) The predicted high accuracy deposition but with internal gaps; (c) MAT path patterns; (d) The predicted void-free deposition but with extra material deposited along the boundary; (e) Adaptive MAT path patterns with varying step-over distance; (f) The predicted void-free deposition with high accuracy at the boundary through using adaptive MAT path [28].

Step-over distance, which is defined as the distance between the next deposition path and the previous one, is always constant for both contour path patterns (refer to **Figure 19a**) and MAT path patterns (refer to **Figure 19c**). For certain geometries, it is not possible to achieve both high accuracy (refer to **Figure 19b**) and void-free (refer to **Figure 19d**) components using paths with constant step-over distance. However, some AM processes, such as wire feed AM process, are capable of producing different widths of deposits within a layer through varying travel speed and wire feed rate, while maintaining a constant deposit height. Therefore, we propose an adaptive path planning strategy that uses continuously varying step-over distances by adjusting the process parameters to deposit beads with variable width within any given path [28]. The developed adaptive MAT path planning algorithm is able to automatically generate path patterns with varying step-over distances (refer to **Figure 19e**) by analyzing geometry information to achieve better part quality (void-free deposition), accuracy at the boundary, and material efficiency, as shown in **Figure 19f**. Example of adaptive MAT path generation for a geometry with multiple holes is shown in **Figure 20**. Examples of void-free additive manufacturing using adaptive MAT paths could be found in [27].

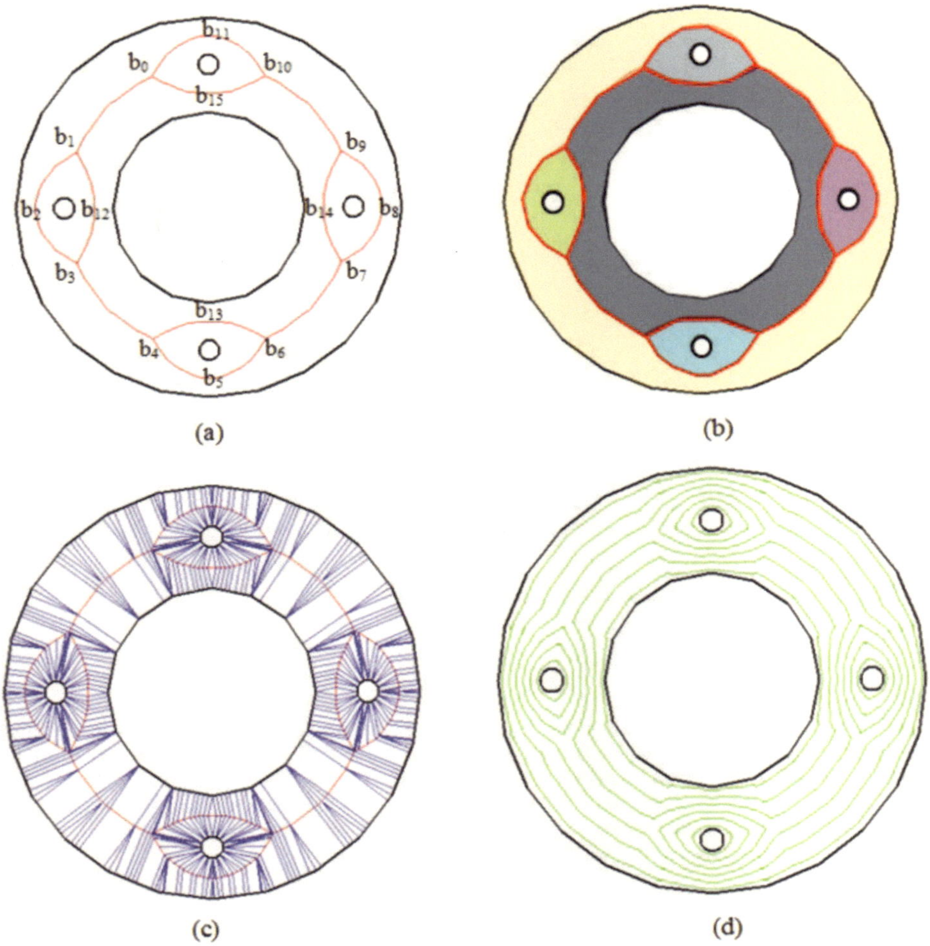

Figure 20. Example of a geometry with multiple holes. (a) Computed branch loops as represented by red lines; (b) Divided six domains as represented by different colors; (c) Generated numerous radiations from the branch points and the various decomposed simple shapes; (d) Generated adaptive paths for the geometry.

14. Summary

This chapter presented slicing strategy for AM, namely unidirection slicing and multidirection slicing. Apart from the slicing, methods of existing path planning, including the newly published MAT-based non-adaptive and adaptive paths, were reviewed for advanced design in AM.

The main conclusions are listed as following:

- Unidirection slicing method is normally used for current commercial AM machines due to its robustness and simplicity. However, supports are required for some complex geometries with overhangs which reduce the accuracy and the building speed of products.
- Multidirection slicing strategy enhances the ability of AM and reduces the usage of supports. Nevertheless, its application is still limited since it requires the AM machine to be able to deposit along multiple directions, and more importantly, algorithms for multidirection slicing are complex and not versatile.
- Ample software are available to generate zigzag, or contour paths for AM process, but the geometries are usually not fully covered by these paths, therefore left the built geometry with voids or gaps. For fabricating the functional products, gap-free paths are beneficial. Therefore, advanced design on path planning is essential as the example of MAT path generation introduced in this chapter.

Author details

Donghong Ding, Zengxi Pan*, Dominic Cuiuri, Huijun Li and Stephen van Duin

*Address all correspondence to: zengxi@uow.edu.au

School of Mechanical, Materials, and Mechatronics Engineering, Faculty of Engineering and Information Science, University of Wollongong, Northfield Ave, Wollongong, NSW, Australia

References

[1] D.-H. Ding, Z.-X. Pan, C. Dominic, and H.-J. Li, "Process planning strategy for wire and arc additive manufacturing," in *Robotic Welding, Intelligence and Automation: RWIA'2014*, T.-J. Tarn, S.-B. Chen, and X.-Q. Chen, Eds., Cham: Springer International Publishing, 2015, pp. 437–450.

[2] S. Choi and K. Kwok, "A tolerant slicing algorithm for layered manufacturing," *Rapid Prototyping Journal*, vol. 8, pp. 161–179, 2002.

[3] D. Ding, Z. Pan, D. Cuiuri, H. Li, N. Larkin, and S. van Duin, "Automatic multi-direction slicing algorithms for wire based additive manufacturing," *Robotics and Computer-Integrated Manufacturing*, vol. 37, pp. 139–150, 2016.

[4] P. Singh and D. Dutta, "Multi-direction slicing for layered manufacturing," *Journal of Computing and Information Science in Engineering*, vol. 1, pp. 129–142, 2001.

[5] Y. Yang, J. Fuh, H. Loh, and Y. Wong, "Multi-orientational deposition to minimize support in the layered manufacturing process," *Journal of Manufacturing Systems*, vol. 22, pp. 116–129, 2003.

[6] J. Ruan, T. E. Sparks, A. Panackal, F. W. Liou, K. Eiamsa-ard, K. Slattery, *et al.*, "Automated slicing for a multiaxis metal deposition system," *Journal of Manufacturing Science and Engineering*, vol. 129, pp. 303–310, 2007.

[7] J. Zhang and F. Liou, "Adaptive slicing for a multi-axis laser aided manufacturing process," *Journal of Mechanical Design*, vol. 126, pp. 254–261, 2004.

[8] P. Singh and D. Dutta, "Offset slices for multidirection layered deposition," *Journal of Manufacturing Science and Engineering*, vol. 130, p. 011011, 2008.

[9] R. Dwivedi and R. Kovacevic, "Process planning for multi-directional laser-based direct metal deposition," *Proceedings of the Institution of Mechanical Engineers, Part C: Journal of Mechanical Engineering Science*, vol. 219, pp. 695–707, 2005.

[10] L. Ren, T. Sparks, J. Ruan, and F. Liou, "Process planning strategies for solid freeform fabrication of metal parts," *Journal of Manufacturing Systems*, vol. 27, pp. 158–165, 2008.

[11] M. R. Dunlavey, "Efficient polygon-filling algorithms for raster displays," *ACM Transactions on Graphics (TOG)*, vol. 2, pp. 264–273, 1983.

[12] S. C. Park and B. K. Choi, "Tool-path planning for direction-parallel area milling," *Computer-Aided Design*, vol. 32, pp. 17–25, 2000.

[13] V. Rajan, V. Srinivasan, and K. A. Tarabanis, "The optimal zigzag direction for filling a two-dimensional region," *Rapid Prototyping Journal*, vol. 7, pp. 231–241, 2001.

[14] R. Farouki, T. Koenig, K. Tarabanis, J. Korein, and J. Batchelder, "Path planning with offset curves for layered fabrication processes," *Journal of Manufacturing Systems*, vol. 14, pp. 355–368, 1995.

[15] Y. Yang, H. Loh, J. Fuh, and Y. Wang, "Equidistant path generation for improving scanning efficiency in layered manufacturing," *Rapid Prototyping Journal*, vol. 8, pp. 30–37, 2002.

[16] H. Li, Z. Dong, and G. W. Vickers, "Optimal toolpath pattern identification for single island, sculptured part rough machining using fuzzy pattern analysis," *Computer-Aided Design*, vol. 26, pp. 787–795, 1994.

[17] F. Ren, Y. Sun, and D. Guo, "Combined reparameterization-based spiral toolpath generation for five-axis sculptured surface machining," *The International Journal of Advanced Manufacturing Technology*, vol. 40, pp. 760–768, 2009.

[18] G. Jin, W. Li, and L. Gao, "An adaptive process planning approach of rapid prototyping and manufacturing," *Robotics and Computer-Integrated Manufacturing*, vol. 29, pp. 23–38, 2013.

[19] H. Wang, P. Jang, and J. A. Stori, "A metric-based approach to two-dimensional (2D) tool-path optimization for high-speed machining," *Transactions-American Society of Mechanical Engineers Journal of Manufacturing Science and Engineering*, vol. 127, p. 33, 2005.

[20] Y. Zhang, Y. Chen, P. Li, and A. T. Male, "Weld deposition-based rapid prototyping: a preliminary study," *Journal of Materials Processing Technology*, vol. 135, pp. 347–357, 2003.

[21] M. Bertoldi, M. Yardimci, C. Pistor, and S. Guceri, "Domain decomposition and space filling curves in toolpath planning and generation," *Proceedings of the 1998 Solid Freeform Fabrication Symposium, The University of Texas at Austin, Austin, Texas*, 1998, pp. 267–74.

[22] T. Wasser, A. D. Jayal, and C. Pistor, "Implementation and evaluation of novel build-styles in fused deposition modeling (FDM)," *Strain*, vol. 5, p. 6, 1999.

[23] R. Dwivedi and R. Kovacevic, "Automated torch path planning using polygon subdivision for solid freeform fabrication based on welding," *Journal of Manufacturing Systems*, vol. 23, pp. 278–291, 2004.

[24] D. Ding, Z. S. Pan, D. Cuiuri, and H. Li, "A tool-path generation strategy for wire and arc additive manufacturing," *The International Journal of Advanced Manufacturing Technology*, vol. 73, pp. 173–183, 2014.

[25] J.-H. Kao and F. B. Prinz, "Optimal motion planning for deposition in layered manufacturing," *Proceedings of DETC*, 1998, pp. 13–16.

[26] D. Ding, Z. Pan, D. Cuiuri, and H. Li, "A practical path planning methodology for wire and arc additive manufacturing of thin-walled structures," *Robotics and Computer-Integrated Manufacturing*, vol. 34, pp. 8–19, 2015.

[27] D. Ding, C. Shen, Z. Pan, D. Cuiuri, H. Li, N. Larkin, *et al.*, "Towards an automated robotic arc-welding-based additive manufacturing system from CAD to finished part," *Computer-Aided Design*, 2016 vol. 73, pp. 66–75.

[28] D. Ding, Z. Pan, D. Cuiuri, H. Li, S. van Duin, and N. Larkin, "Bead modelling and implementation of adaptive MAT path in wire and arc additive manufacturing," *Robotics and Computer-Integrated Manufacturing*, vol. 39, pp. 32–42, 2016.

Chapter 2

Color 3D Printing: Theory, Method, and Application

Guangxue Chen, Chen Chen, Zhaohui Yu, Hao Yin,
Liuxi He and Jiangping Yuan

Additional information is available at the end of the chapter

http://dx.doi.org/10.5772/63944

Abstract

Our research team proposes a colored manufacturing technology with a layer-by-layer printing process. Using digital inkjet printing in layer-by-layer printing color graphics, a further low-cost color 3D Printing (3DP) technology can be developed. It can provide an integrated way to prototype and reproduce 3D objects, from concept to design and manufacturing. Ultimately, with fast graphics printing method, it guarantees a feasible way to further promote cultural and creative products.

Keywords: 3D Printing, color reproduction, historic preservation, topographic map, 3D geographical model

1. Introduction

Recently, the application of personalized 3D Printing (3DP) in culture and art has been experiencing a gradual increase in demand. Beautiful lively items with exquisite appearance are in need of fabrication and reproduction, with high artistic and economic value, which is greatly expanding in the market of rapid prototyping.

However, the 3D Printing process has restrictions due to forming materials and equipment; the presence of monochrome objects makes it difficult to meet the demand of high-quality reproduction. These limit the further development and prosperity of the emerging 3D market with high value-added potential products in the field of culture and art.

This chapter presents the technique and configuration needed to develop a workflow from 3D digital model to rendering images and color 3D Printing. The chapter is divided into four parts.

The first part reviews the current situation in terms of 3D Printing and focuses on color-related research and applications.

At first, a quick discussion is presented about models, which are constructed out of vertices and polygons, and the impact of modeling techniques in terms of mapping color information into model itself. And some formats as examples are used to demonstrate the differences.

The second part shows the theory of color 3D Printing process, at first about the work of our research and focuses on the hardware and process to print the interpreted data into objects and transfer color information into the model. The related 2D printing applications as well as a series of new issues are also presented.

The third part shows the modeling data representation and data transformation with a short review of the rendering pipeline for color 3D Printing.

This part provides specific mechanics from 3D modeling to "ready to print." A graphics processor usually performs several pervertex calculations in addition to the transformation from object space to window space. This specifies how the fragment data are used to determine the final color and the final depth of each pixel.

The last part demonstrates a few instances of color objects produced with the process, showing the color reproduction and further potential applications.

2. Current color printing processes

3D Printing technology is a powerful technology that is leading significant changes in many areas. 3D Printing can be defined as a process creating an object by printing each sliced section layer by layer until finishing a completed object. In traditional manufacturing industries, more customized and creative functions could be provided By 3D Printing [1]. Currently, the 3D Printing is still in the research era of printing precision and speed. However, this is also the first step to exploit all advanced 3D Printing techniques, such as whole colorful printing and large-sized integrated printing. Moreover, color 3D Printing has become a hot topic in the 3D Printing field for many researchers and manufacturers. Even though the colorimetry is not necessary for all reproductions of 3D color objects, but this function is urged to provide and improve for manufacturers.

From the view of classic materials and applications, the 3D Printing process can be divided into six categories: powder-based 3D Printing, plastic-based 3D Printing, paper-based 3D Printing, organism-based 3D Printing, food-based 3D Printing, and metal-based 3D Printing. In addition, the glass-based 3D Printing is a new useful process to produce amazing glass artistic wok which only focused on its transparency and styling. Colorimetry is the core feature of color objects, showing its artistry and aesthetics, which should be reproduced accurately to meet the advanced demands. Recently, many published articles have shown various researches about color 3D Printing from different areas. According to 3D Printing applications, generic color 3D Printing technologies are demonstrated with relative principles and instances in the following parts.

2.1. Powder-based color 3D Printing

Powder-based color 3D Printing is widely applied into artistic work, geographic information system, and Medical buttress applications [4]. Based on powder, two 3D Printing processes can be found in manufactured instances. One is the famous 3DP process that is innovated and patented by Massachusetts Institute of Technology. The 3DP process is mainly using basic powder and color blinder combined by the inkjet printing method to build a color part. Another is the selective laser sintering process (SLS) which creates a completed model by sintering each Powder layer [3]. The big advantage of powder is its good form ability and stability. At the same time, some special powder also provides a good connection with color binder based on low temperature.

The 3DP process was licensed to Z-Corporation Company, focusing on the concept model implementation, which developed and sold the first high-definition color 3D printer named Spectrum Z510 in 2015. The colorimetry features of color objects are reproduced by inkjetting various color binders on powder layers. Three current color binders can be obtained, such as primary cyan, magenta, and yellow. The surface color of objects can be presented by the mixture of various ratios of above primary color binders. However, the black ink in the traditional inkjet printing is replaced by clear binder that is connected to most noncolor powder layers. That is, the ingredients of color binders play an important role for the reproduction of color 3D objects. For this 3D process, it easily achieves the postfinished step by using the designed infiltrates to make a better color performance. These finished methods were introduced and contrasted by Maja Stanic using two different Infiltrates in 2012 [21]. In addition, to infiltrates, the wax is also a good supplement material for the fashion designers to improve the color features of 3D printed artifacts under this process. The 3DP process is a good technology to build Buddha state with a bit heavy.

According to the current 3D printer, the SLS process can be achieved by two general approaches: direct method and indirect method [5]. The basic powder is varied from the metal powder to plastic powder to calcium powder in diverse applications. In this case, the color features of productions of the SLS process are limited to the simple or single color. From the current device and application, this depends on the original color of basic powder. For the plastic powder, the color of the printed objects nearly consists the original color of the basic powder. For the calcium powder, the color of 3D printed products may be changed in the lightness. Moreover, it is easy to produce different primary color powders by material manufacturers. For the last, the color of mental powder object can be added to metallic gloss. The SLS process provides a useful approach to create tissue scaffolds such as Bone and teeth [8].

2.2. Plastic-based color 3D Printing

Plastic-based color 3D Printing is a cheap and fast process to provide printed color objects, which widely applied in engineering parts. Based on the plastic material or the photosensitive resin, stereo lithography appearance (SLA) and fused deposition modeling (FDM) are two representative 3D Printing technologies. This method makes the most of the three features of plastic or resin such as easy fusion, good formality, and stable irreversibility. Moreover, the plastic-based color 3D Printing had captured a big market share of 3D Printing industry. There are two factors, such as physical strength and color characteristics, to make a huge success.

The SLA process is based on the liquid resin and the ultraviolet laser; some famous research institutes show great enthusiasm into the color function. For example, America Materialise Company had used two resins (red and green) to print a medical model to distinguish the diseased tissue and normal tissue to diagnostic etiology. University of Tokyo had adapted the inkjet printer to achieve the surface colorization for each designed layer in the SLA procedure. Pusan National University had presented a colored MEMS visualization model and layer interconnection visualization of electronic packaging part by using an inner visible multicolor Prototyping process [9]. Their idea is that first the resin in the noncolor area of each layer is solidified and then the resin in the color area is replaced by a new color resin to form the color part until A completed object [7].

The FDM process is patented and advertised by Stratasys Company, which used different colored plastic silks with low melting point to form a new color object. For FDM printer, it usually contains four hot-melt nozzles to provide three primary colors and one model bracket. The Fortus 900mc printer is an amazing star product which provides super print size with nine printing substrates. In China, Tsinghua university had developed a new own FDM printer without the laser melting device which sharply decreased the price of this printer. For this process, hundreds of colors can be produced in the surface of different color objects. However, it is so difficult to make gradient color in the local area of The whole object [6].

2.3. Paper-based color 3D Printing

The paper-based color 3D Printing is a technology using regular office paper as the main substrate and adding color inks to implement Thousands of colors [10]. This process is similar to inkjet printing but extend into three dimensional formation. Generally, according to forming principle, the paper-based color 3D Printing technology also contains two processes implanted by 3D printers such as MCOR IRIS series 3D printer of Mcor-technology Company and Mimaki UJF-3042 UV printer of Mimaki Company. For these two processes, they both provide tempting color reproduction of nonlarge color objects. Currently, the MCOR IRIS series 3D printer can achieve 256 × 169 × 150 mm (for an A4 paper) and 238.5 × 175 × 149.9 mm (for a letter-sized paper). However, the Mimaki UJF-3042 UV printer can build 300 × 420 × 50 mm. In addition, their colorization principles are also different even though using the basic office paper.

The MCOR IRIS series 3D printer is an undoubted powerful color 3D printer that is frequently applied into creative cultural industry. This process is divided into two parts such as colorization and modeling. The first step of colorization is to print color inks within the contour area at both sides of each paper to make sure the ink fills sufficiently the inner space of the paper. When one layer is colorized, it is sent for trimming and gluing. The color of cutting cross section and the surface paper build the completed surface of the color object. For this printer, the printing speed is and the build size is also limited. To solve the first issue, JP Yuan and GX Chen proposed a speedup method for paper-based 3D color printing by developing a new fast and reliable coloration algorithm using geometric information in STL file, combining the existing different slicing methods, which can be applied in the paper-based color 3D printer based on The uniform slicing thickness [11]. To solve the second issue, JP Yuan and GX Chen presented a feasible visualization of large-size 3D color models which are beyond the maximum print size of the newest paper-based 3D printer using 3D cutting-bonding frame (3D-CBF), and evaluated their effects of cutting angle and layout method on the printing time of The designed models [12].

The Mimaki UJF-3042 UV printer is used as the tiny 3D Printing or 2.5D printing tasks in which the height of the color object is small but it requires high resolution. This process is similar to the traditional inkjet printing using color UV inks. It adapted six UV ink boxes and nozzles such as CMYK four-color inks and two transparent inks. A major difference is that each layer can deposit on the previous layer until finishing a whole object with excellent color surface. C. Chen et al. analyzed the effect of colorization sequences of color ink and white ink on oil painting 3D reproduction and found that the color ink printed in the upper layers can form an effective smooth and Continuous curved surface [13]. H.M. Wang and G.X. Chen also studied the process applied in the Taiwan map 3D reproduction based on different substrates including office paper, and provided a printed sequence frame to retard the jagged Edge phenomenon [14, 22].

2.4. Cell-based color 3D Printing

Cell-based color 3D Printing is popular in medical applications such as the organism printing and tissue producing. From the printing principle, the cell-based color 3D Printing is similar to the inkjet printing, whereas the basic material is not ink but various cells. The color of cell in body behaved like a natural color, and this color is stable even if it maybe changed with the oxygen concentration In regular direction [2]. For the outside soft tissue of body, the color of prostheses is very important for transplant recipients. Xiao et al. had developed a 3D color image reproduction system for additive manufacturing of facial prostheses based on the ZP510 Printer [15]. This frame provided a good instance and flowchart for the color reproduction of 3D Printing cell. C.L. Cheung et al. also introduced the development and face validation of a pediatric pyeloplasty simulator using a low-cost laparoscopic drylaboratory model developed with three-dimensional (3D) printing and silicone modeling and provided the color reproduction application of kidney Based on silicone rubber[16].

2.5. Food-based color 3D Printing

Food-based color 3D Printing is a generalized definition which contains all processes that are achieved by 3D Printing in food industry. For the personalized food, the color is also an important factor to attract customer and provide appetite temptation. Actually, the snacks are available and suitable for 3D Printing. Yang et al. had used sugar and different flavor binders to fabricate complex sculptural cakes For special events [17]. Foodjet Company had used pneumatic membrane nozzle jets to deposit selected material drops onto pizza bases, biscuits, and cupcakes by the Inkjet printing process [18]. Fu Jia et al. investigated the feasibility of supply-chain-centric business models in 3D chocolate printing and provided suggestions for color requirements of commercial chocolate produced By the food 3D printer [19]. The above instances provided color surface but just simple pure color. In order to achieve more diverse and colorful eatable food, Jie Sun et al. introduced and analyzed the effect of multimaterials and multiprintheads on color food products, and provided a feasible method to optimize the reproduction of color Eatable food [20].

2.6. Current researches of color 3D Printing

3D color printing is a promising technology that will produce all the color objects in daily life; however, there are rare relevant researches now. For the different 3D color printing processes, the colorization principles are varied from the devices and software. Like the printing graphic, the color reproduction of color 3D Printing is also a key issue for manufacturers and designers. So, 3D color measurement, 3D color characterization, and 3D color quality evaluation are the three key factors to improve the color reproduction of 3D color objects.

For the 3D color measurement, Stanic et al. systematically introduced the selected measurement method based on inkjet printing parts. This method is similar to CIE guide to the color measurement of printing graphics. In color 3D Printing instances, some researches proposed their measurement devices and conditions. Stanic had adopted the GretagMacbeth (X-rite) XTH sphere spectrophotometer as the measurement for the printed powder-based color object. Xiao et al. used a Minolta CM-2600d spectrophotometer using Spectra Magic NX Color Data Software that was employed to study color measurements in CIELAB values. The illuminant was based on the CIE standard D65 to simulate skin color in daylight conditions. He et al. selected the X-rite 530 spectrophotometer for the 3D color patches printed by the Mimaki UJF-3042 UV printer. Obviously, the color measurement tool is the same as that in printing graphics. This maybe caused the unexpected color error under the curved surface.

3. Color 3D Printing

Whether in personalized design work, advanced aerospace component manufacturing, or even in three-dimensional cells, the color and appearance issues cannot be ignored. Although in industrial design, mechanical design, or creative and cultural industries, there are not yet satisfactory color reproduction solutions or a low-cost color reproduction technology. But 3D color printing already has a huge and significant impact on life and society. Along with the development of different three-dimensional manufacturing processes, the 3D color printing will also accumulate certain color reproduction techniques.

3D color characterization is an important tool to communicate the color reproduction. Currently, the uniform color spaces are also applied in 3D color characterization, such as the CIELab system. Based on the existing color space, the color value is easily measured and computed with current devices and metrics. When color data were obtained, the color gamut of printed device, the corresponding substrate, and colorant can be described and compared. This can be easily obtained by the Profilemaker software. DE2000 and DEcmc color difference values were widely used as the contrast of color difference.

Color quality has an irreplaceable effect on the surface appearance of the color objects. Color properties and stability are the two key factors that influence the product quality. Stanic et al.

http://dx.doi.org/10.5772/63944

had studied the effect of lightfastness experiment on color difference of powder-based 3D color printing objects. He had discussed the effect of aqueous adhesives on color of paper surface in the paper-based 3D Printing. Actually, the process parameters and basic material can achieve the color quality without a fixed trend. At the same time, the postfinished methods can affect the color quality. Carinna Parraman had presented the effect of wax and selected color patches on the color appearance of color patches even if the influence mechanism is under the water.

3.1. Stages of development

The evolution of 3D Printing technology and materials for inkjet printing and the development of 3D color printing can be divided into the following three stages:

(1) Single color: These stages are originated by Charles Hull, America. In 1984, he invented the SLA process and established the 3D Systems Company in 1986. In 1987, 3D Systems manufactured its first 3D printer called SLA250. The printer had no substantial breakthrough in terms of color. This stage is also known as "black and white" 3D Printing manufacturing. For example, liquid photosensitive resin material used by this printer itself has some specific colors, after stereo lithography has been finished, the product will be followed with a single basic monotone color. This stage is more dependent on post-processing, such as painting, dipping, and so on, to make products more colorful.

(2) Discrete color printing: This stage of technique turned to become mature in 2000. Multijet printing technology delivers two or three kinds of plastic in filament form with a variety of colors. It can be colored before or after printing. 3D color printers currently dominate the market and are mostly based on this technology. Discrete color printing can surely achieve certain kinds of appearance for products. There are already roughly 160 kinds of synthetic colors available, but the fineness of the results is not satisfactory, many Chinese 3D printing device vendors already mastered this colored plastic filaments fusion technology.

(3) Continuous tone printing: This stage originated in 2005, marked as the first high-definition color 3D printer by the Z Corp Spectrum Z510. The 3D printer is capable of printing high-resolution colorful 3D products. It features a direct prototyping process to achieve continuous color reproduction. After completion of forming the 3D printed models, the corresponding colors can show better results without post-treatments. This continuous tone color technology can be achieved by a variety of printing processes, including adhesive binding powder material, colorful liquid UV-resin-based mixing or paper-based subtractive printing.

3.2. Gamut characteristics with 3D models

In gamut mapping, the image gamut with a certain range of colors can be expressed. Then, it can be seen as a subset of the larger color space. The gamut mapping method based on image content is an important direction for research currently. Gamut characteristics of image determine the gamut mapping source, which is the mean issue to be considered at first as color gamut mapping is carried out based on the image content. The chromaticity of a color image is one of the input values for gamut mapping and a main parameter for the evaluation and calculation of color reproduction. Thus, a digital image can be analyzed following the steps outlined in **Figure 1**. And, then an appropriate gamut mapping algorithm can be selected for specific digital image gamut characteristics.

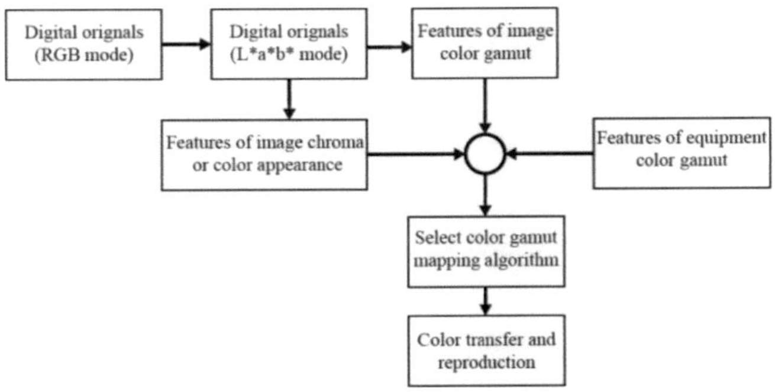

Figure 1. Digital image analyzing process.

3.3. 3D Printing color management

Consider the Z510 3D printer of ZCORP Company for example, during the 3D color inkjet printing, coloring is achieved by adding CMY or CMYK color adhesive to mixed transparent adhesives. The colorant in adhesive has its main role as coloring agent. 3D digital models can have a variety of coloring methods, such as directly on the surface, or coloration is applied as the image content, in frequently used information carrier, such as TIFF JPEG file formats files. Of course, a colorful 3D digital model need to be saved in formats that can include its color information, such as VRML, PLY, ZPR, or other proprietary formats.

Issues of coloring of 3D inkjet printing include 3D color reproduction, color consistency and control, color gamut range, effects of materials and processes, the surface characteristics of 3D Printing, and color measurement and other urgent problems to solve. Another related problem is the persistence and firmness of the finished 3D model; both are becoming increasingly important in the 3D inkjet printing and other common applications such as space planning and art and design applications.

3.4. Color detection, measurement, and control

It has been proposed for 10 years that rapid prototyping technology can produce colored objects with basic color or full spectrum of colors, such as the Zcrop Company developed the 3D inkjet printing system. While the conventional color inkjet printing is being researched and developed for a long period of time, 3D inkjet technology prints different materials with the corresponding coloring processes, which may have special problems. The overall performance is affected by the physical appearance of the object and color appearance, material's gamut, liquidity of base material before forming, the positioning accuracy of printed surface, and the accuracy of scanned digital information. The main effect on finished surface of 3D Printing objects is caused by different printing principles, printing equipment, as well as different features and rough texture of powdered print material particles.

Different positions on the surface of printed objects and different x, y, and z location coordinates on 3D model will affect the result of colored reproduction. This is due to manufacturing method and layered approach to generate model, which causes difficulties in superimposition of layers to achieve true vertical surface. The results have curved surfaces with certain angles, so it inevitably has questions of aliasing and grinning. Of course, parts of the surface effect issues can be solved during subsequent processing.

Since the finished surface has inherent optical, chemical, and physical characteristics, the finishing coloring processes provide different results. **Figure 2** shows print sample color test chart and their contrasts of saturation and lightness characteristics under three different conditions. The left sample was impregnated with cyanoacrylate, the middle sample was blank, and the right one was impregnated with epoxy resin.

Figure 2. 3D Printing color test charts.

Another significant issue not yet tackled is that there is no precise equipment to measure color surface structure and characteristics of 3D Printing product. Currently, we can use a spectrophotometer with an 8-degree field of view to measure them. Spectrophotometer is widely used in textiles, dyes, paints, plastics, and automotive industry as it may require any kind of surface texture and gloss color measurement.

In order to describe 3D Printing color reproduction capability, and monitor and control color output on a 3D printer, we need professional precise testing instruments and software to obtain real sample color and test strip. The color test strip can be generic, such as Zcrop icon's color test bar or special test strips can be developed. Those special strips should be designed specifically to measure appearance and size, while a good variety of colors characteristic can be expressed. In order to facilitate the existing spectrophotometric measurement, color test strips are preferably designed by flat form. **Figure 3** shows some specific designs and printouts of color test strips.

Figure 3. 3D Printing color test strips.

Of course, the color test strips can also be of 3D print themselves, which is more suited to some particular measurement requirement, for example, to observe and monitor quality of color reproduction on flat structure, or curve or groove surface. Color test strips can be considered as a kind of color authenticity visual aids to help users predict the printing process. They can also be considered as a direct tool like 3D color swatch book for user to select the desired color.

3.5. Color stability and durability

3D Printing applications are being used in new areas, people discuss about a wide range of common properties, such as durability and stability of color 3D Printing objects. These issues are of more interest to art and design people, who use rapid prototyping technology to aid their work, because they publicly display their works of art, and geospatial information systems, architecture, space planning, scenic modeling, and urban design need excellent color effects. Color is a core part of printed objects, and people expect the original effects to remain for a certain time.

Duration and stability depend on several factors, such as light, humidity, temperature, and the amount of air in contact. Graphics, printing industry, and color science and other industries have been developed and used a lot of color standards and standard printing methods developed to study and predict the color durability and stability. This is more significant in inkjet printing in commercial and art fields. Whether 3D Printing is also suitable for the above-mentioned research methodology, and the materials, air exposure and other factors can influence the finished colors aging mechanism are real issues focused by current researches.

Currently, the durability and stability study of 3D printed color focus on how to find the right way to simulate aging or change real 3D prints, which are exposing in different environment conditions. For example, different lighting, temperature, humidity, the finished color of 3D Printing, and the behavior of color change with different impregnating agent can be simulated and predicted. Custom color strips should be developed in test researches, as well as corresponding colors and test procedures, so that people can conduct relevant research.

http://dx.doi.org/10.5772/63944

3.6. Digital coloring method of 3D Printing

A color model describes the displayed color effects of print. Different color models define different range of colors. The fields of their applications are different. These applications are in addition to determining the number of color inside a model, the number of color channels, and the file size of color image information. Here we present several related concepts of color images.

Bit depth: Bit depth, also called the pixel depth or the color depth, is used to measure color information in the image to display or print. Greater bit depth means that the digital image has more colors and more accurate color representation.

Gamut: This represents the range of colors capable of displaying or printed. Lab color space has the widest color gamut, which can contain all the color types from RGB and CMYK color mode. Typically, RGB color gamut can be displayed on a computer monitor or a television screen. However, some, such as pure cyan or pure yellow and other colors, cannot be accurately displayed on digital monitors. CMYK color space has narrower gamut, containing only color types used in printing, which can be printed with ink. When a color cannot be printed or displayed on the screen, it is called out of the gamut, that is, beyond the CMYK color gamut.

Color channel: Each color image has one or more channels, and each channel can store the color information of imagery elements. Default color channels in an image depend on their color space. The CMYK image has at least four channels, each representing the color layers cyan, magenta, yellow, and black. The RGB image has three channels, representing the color layers of red, green, and blue.

Gray image: This type of image can exhibit a rich range of tone. It uses up to 256 shades of gray. Each pixel has a grayscale image luminance value from 0 (black) to 255 (white). Black and white or grayscale scanning device produces images that often appear in this gray image. If a high-quality color image is converted into a black and white image, Photoshop will display all the color information in the original image. When a gray image is converted into an RGB image, the color values of pixels will be lower than the previous gray image values. A gray image can be converted to a Lab or a CMYK color image.

3.7. Two-dimensional interpolation method

Colorful images cannot scale without interpolation methods, and commonly used color interpolation methods are bilinear interpolation, cubic convolution, and B-spline interpolation.

3.7.1. Bilinear interpolation

As shown in **Figure 4(a)** and **(b)**, at R and B positions, the color component of G is to be interpolated; this approach uses current G values of vertically and horizontally adjacent four pixels; these four values are added and an average is taken, which is set as G interpolation value of this position. Position B uses the average of adjacent four G values. As shown in **Figure 4(c)** and **(d)**, if the G value is to be inserted inside the R and B color components, and this G position is vertically or horizontally adjacent to R and B, then the sum of vertical or horizontal two values of the same color can be used and an average is taken. This is called a color component interpolation method.

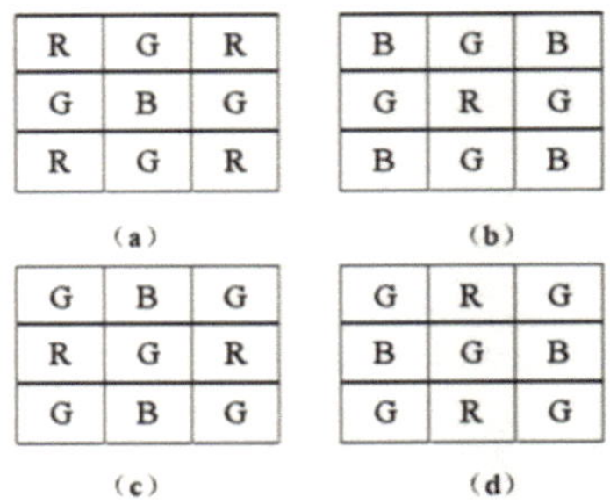

Figure 4. Bilinear interpolation.

3.7.2. Cubic convolution

A cubic convolution interpolation method is one of the most commonly used functions within the grid data interpolation method. It uses the pixel to be estimated from the nearest 16-pixel value, and the result is calculated according to the formula of cubic convolution. As shown in **Figure 5**, there are many cubic convolution formulas; some produce low-pass filtering effects and some produce a high-pass filter effect. A better approach would be to achieve a balance in the tradeoffs between high and low frequency information. Meanwhile, the bandwidth of cubic convolution is wider than bilinear interpolation, and in the image of high frequency, cubic convolution shows better interpolation results than bilinear interpolation.

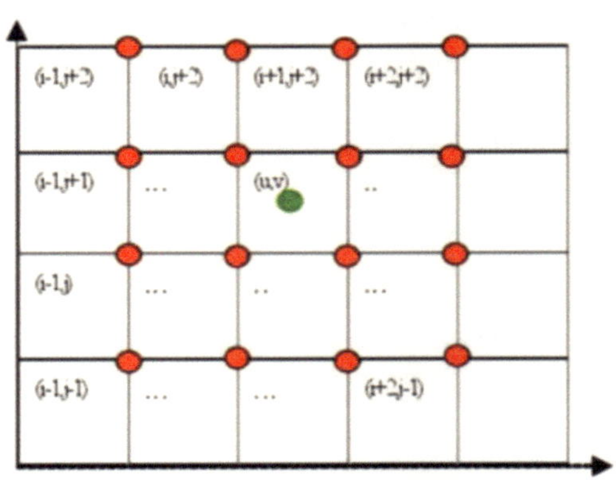

Figure 5. Cubic convolution interpolation.

The principle of cubic convolution interpolation is based on the interpolation theory, which shows that the cubic convolution of two-dimensional image with generally equally spaced sampling can be expressed as

$$s(x,y)=\sum_l\sum_m c(x_l,y_m)\times h(|x-x_1|)\times h(|y-y_m|) \quad (1)$$

where h is referred to as the interpolation weight function, $s(x, y)$ is the results of unknown interpolation points, and sample $c(xl, ym)$ represents the pixel (xl, ym). Obviously, the selection of interpolation function h is the key of any convolution interpolation method. The weight function is a cubic interval function, and the nature of its characteristics can be assessed by the range of its frequency sections. An ideal weight function has unity gain in the band-pass range, has a zero gain in the band-stop range, and thus the signal components with different frequencies can be effectively suppressed.

3.7.3. B-spline interpolation

Commonly known B-spline interpolation methods use interpolation function with discrete data to generate continuous function, and then the function can resample and thereby generate new interpolation values. A continuous function can be generated with the following formula:

$$f(x)=\sum_{k\in s} f_k\varphi_{\text{int}}(x-k) \qquad \forall_x \in R \quad (2)$$

In the above formula, fk is the known discrete data sequence, and ϕ int(x) is the interpolation function. In order to ensure the accuracy of interpolation, usually it requires ϕ int(x) values of 1 at the origin of coordinates, and in other integer coordinate points, its value is 0, for example, one of the ideal interpolation function is sin(x).

3.8. Three-dimensional real volume rendering

A solid color has two types of applications: one is coloring the surface of 3D solid objects. However, sometimes in order to meet industry needs, products internal parts can have color too. For example, if each part in a car engine has its own color code, it will provide the product management, assembly workers, and maintenance workers great convenience. The other type is coloring the entire inner side of an object or add color to a three-dimensional entity internally which is a commonly used method of volume rendering.

The goal of drawing color information in two-dimensional surface graphics is to add color on the face, same as the concept of pixels in plane. The number of pixels per square inch defines the concept of resolution. Corresponding to surface graphics, there is volume graphics and its models. In the application of three-dimensional grid to directly draw a three-dimensional scene, each grid in space corresponds to a value of particular property, which can be measured.

Those three-dimensional grids are called voxel. Voxel can be seen as a two-dimensional pixel in three-dimensional space. It can be understood that an entity is composed of an amount of small cubes or other three-dimensional primaries. An entity may be a function of an image but composed by voxels. In real three-dimensional scene visualization, the information of a three-dimensional coordinates is corresponding to attribute data set. In this real visualization, the three-dimensional coordinates (x, y, z) are expressed as dependent arguments to space entity for object geometry modeling. In that way, an established model can not only achieve true three-dimensional visualization, but it can also be used in a three-dimensional spatial analysis.

In general, three-dimensional spatial data are continuous, and numerical results or measurement data are discrete and are the results of consecutive sampling of real three-dimensional scene. A volume-rendering technique directly samples this three-dimensional space into two-dimensional image on the screen as accurately as possible to represent the original 3D data. The two-dimensional image on the screen depends on the frame buffer corresponding to the brightness value of each pixel, which is a two-dimensional discrete data set.

Thus, the real volume rendering transforms the three-dimensional spatial discrete data into two-dimensional discrete lattice. If discrete three-dimensional data set can be converted into discrete two-dimensional data matrix, first it must be resampled into three-dimensional data set. Second, the contribution of each voxel in three-dimensional space for final two-dimensional image data should be considered so that the image should be composted. Therefore, real volume rendering of three-dimensional model is a process of resampling discrete data set and image composting. Resampling in theory should have the following steps:

(1) Select the appropriate reconstruction function, three-dimensional data set for discrete three-dimensional convolution, and continuous three-dimensional reconstruction of the data field.

(2) Geometric transformations of three-dimensional continuous data set should be based on a given observation direction.

(3) Since sampling point with screen resolution is given, the sampled signal's Nyquist frequency limit can be calculated, and use low-pass filter function to remove frequency components above this limit.

(4) Resample the filtered function.

The three-dimensional convolution calculation is very time-consuming; in the past few years, a number of scholars have proposed several rendering algorithms to be implemented, mostly discrete methods. Most noticeable is proposed by Westover, a parallel, feed-forward volume-rendering algorithm (i.e., a unit projection method), and M. Levoy presented volume-rendering algorithm with image space order (i.e., a ray casting method).

The conversion brings brightness values from each sampled point into an image plane, that is, to calculate the contribution of brightness values from each sampled point and then a final image can be composed. The idea is first determine the projected order of voxels and then break voxels into several smaller voxels so that these smaller voxels can only be seen from a positive side or a reverse side from the projection direction. Positive and reverse side surfaces are corresponding to projection surface with a projected polygon, so, in the full use of its two-dimensional projection coherence, polygon can be interpolated with a cumulative value of light intensity and transparency. The process is shown in **Figure 6**.

http://dx.doi.org/10.5772/63944

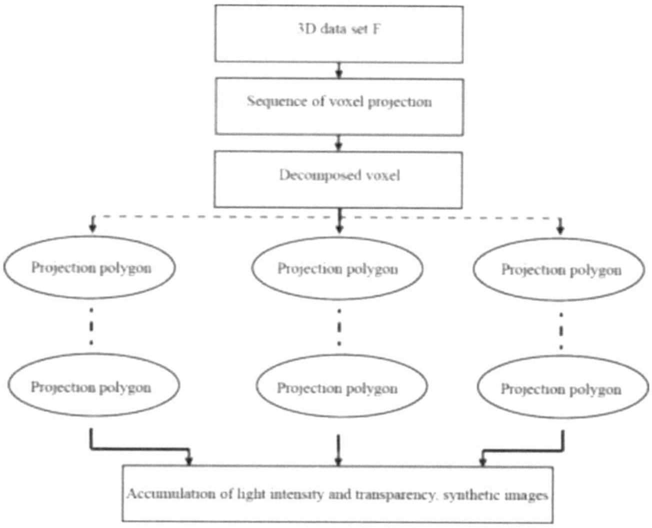

Figure 6. Convolution calculation process.

A ray casting method is also known as ray tracing method. It follows the following basic steps: start from each pixel on the screen, according to the current position of the viewpoint, launch a line through the data set, select a range of sampling points along the line, and use the interpolation method to calculate the value of the opacity points and colored sampling points. Then, from the front to the rear or from the rear to the front sequence to composite color based on their corresponding color value and opacity value at each sampling point, the screen color of each pixel is obtained. The process is shown in **Figure 7**.

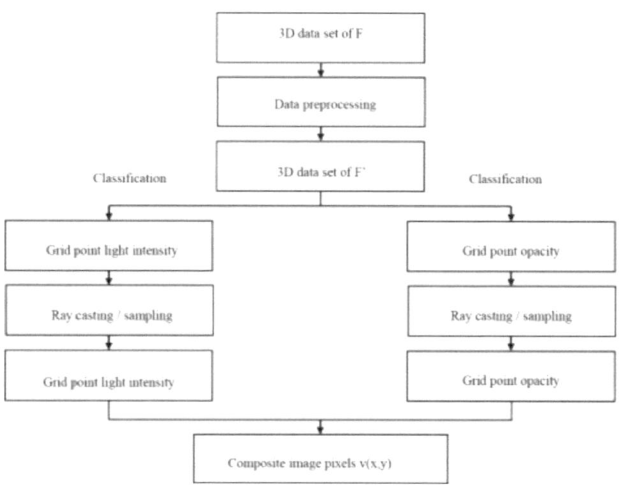

Figure 7. Ray casting/sampling process.

3.8.1. Gradient discrete solid coloring method

There are two kinds of algorithms in 3D solid drawing; ray casting contains intersection operations of light and solid units. For a larger number of units in finite-element data set, intersection calculation is very time-consuming. Because the unit projection method requires advance knowledge of attributes to determine unit order and unit projection, this unit sorting is also very time-consuming. Because two ways of volume rendering are costly and time-consuming, associated overhead and space operation usage are unusually large, up to the time complexity *O*(N3), so it is very complicated to achieve the right color gradient on each unit voxel. Therefore, this section proposes a new method of coloring the internal entities—gradient discrete solid coloring method. By this method, the coloring process of three-dimensional solid can be reduced to a two-dimensional coloring process; its flexibility and efficiency have been greatly improved and the most important characteristic is that it can achieve the right color gradient effect.

Internal coloring process of three-dimensional solid can be completed by adding color information to each discrete layer of the solid. Since each discrete layer shows color information, the solid can obtain the color features when layers superimpose together. A gradient discrete solid coloring method is shown in **Figure 8**.

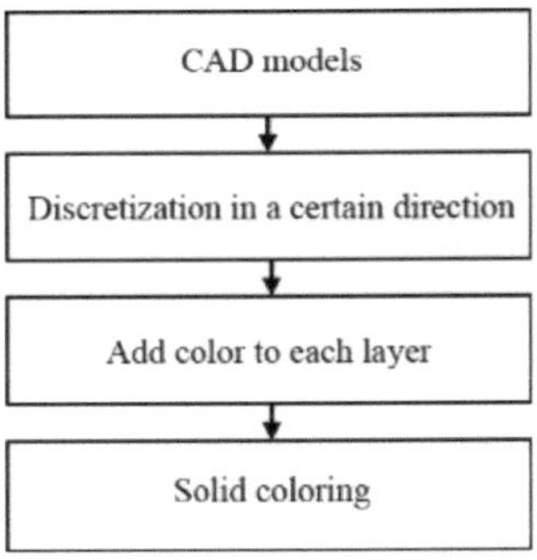

Figure 8. Gradient discrete solid coloring process.

Discrete three-dimensional modeling in rapid prototyping technology uses discrete method to easily and quickly finish coloring. User would like to see as much of the color information by adding color to objects, or can take the initiative to develop new color with more interactive operations, which is desired by the user. The smooth and soft effect based on the human eye make user wants to add gradient colors.

However, as the human eye can only recognize up to about 2000 types of colors, so computer color representation is not strictly continuous and generally uses only eight binary digits to represent R (8 bit), G (8 bit), B (8 bit) of colors, but for the color change, it is often represented by R, G, B three-color superposition:

$$C(R,G,B) = F_R(R) + F_G(G) + F_B(B) \tag{3}$$

where F function is used to obtain the channel points for one pixel. While each color is represented by only 8-bit value; however, it fully matches the amount of human eye distinguishable colors.

3.8.2. Color printing forming process

Three-dimensional rapid prototyping and other 3D Printing all have slicing process. One appropriate direction as the slice direction must be selected, and from one end to the other end of the solid, a series of successive sections S1, S2 … Si … Sn–1, Sn can be given. Note C represents the color information of each slice so that S1(C1), S2(C2), S3(C3), … Si(Ci) … Sn(Cn) is the color information for each slice plane.

According to the differences in range of different colors recognizable by human eye, limited critical points between (i, j) are inserted so that a good transition from Ci to Cj can be achieved; in fact, in order to adapt the characteristics of 3D Printing, these critical points can be taken on the axis of the direction of slicing and the critical points are in the slice plane; and each slice corresponds to a critical point, and the color of this point is attached to the entire slice layer. For a layer-by-layer approach of 3D Printing and rapid prototyping technology, there are same thickness layer methods, adaptive hierarchical methods, but the interpolation methods are not limited to slicing methods. There are three kinds of color interpolation methods, namely, cosine color gradients method, cosine adjustable color gradient method, and power gradient method.

When a three-dimensional printer selects an appropriate image file format as its original before coloring, in the RGB mode the color image pixel is displayed; after the conversion of the image from the RGB mode to the CMYK mode, each pixel in the RGB mode will be transferred to a corresponding CMYK mode matrix for nozzle to read, where one pixel corresponds to a 3 × 3 or 4 × 4 or an $N \times N$ matrix. Print heads receive C, M, Y, K data, and when the jet is connected with the color binder, during printing, each nozzle receives "0" or "1" for control nozzle spray or without spray.

Machine uses lattice format to represent layer information. At every move, the nozzle will be given a forming region with fixed height and a variable width. The graphics are divided into several smaller areas for printing; inkjet molding operation is actually a loop operation, wherein each loop finishes a certain area of a small area and deals with them accordingly. Image information of each layer at first is loaded into the buffer buf[x, y], which is an array of small blocks of image information, known as graphics files. It should be noted that the number of points in the array buf is not the number of points required for the actual nozzle, buf only provides pixel values of the image, and it needs to go through a particular image transfer so that it can become a nozzle-forming image corresponding to a pixel matrix. If we use a 3 × 3 lattice, a pixel in buf corresponds to nine points of nozzle point data; in accordance with such corresponding relation, an image forming process is completed for each small area.

After completion of one layer of material, the printing nozzles moves back to the origin, and platform falls to a certain height in the Z axle, such as 0.5 mm. The printing nozzles restart to form the next layer. This is called one layer forming cycle, and after certain cycles, the color solid is finally completed.

3.8.3. Coloring module software development

Three-dimensional color printer with solid forming has one key technology, that is, the development of coloring software modules. The software can process STL file data, form a solid layer contour, and add an internal solid modeling tool to add gradient color information. According to the requirements of colored objects, a coloring module should have two functions:

1. *STL file data processing function*: (a) read and parse STL files; (b) establish triangular facet topology; (c) design solid slicing algorithm and contour generation.

2. *Gradient interpolation coloring on each layer*: (a) linear interpolation coloring; (b) cosine interpolation coloring; (c) cosine G color tone interpolation coloring; (d) power interpolation coloring.

Using UGNX software or CAD drawing software to make three-dimensional solid modeling, mapping is performed through the menu options after forming "* .stl" file. With common 3D modeling software, in the "File" menu there are "Export" command, and it can export a "* .stl" file, and then manually change it into ".txt" type files. To understand the file format, use notepad to open "* .txt", and then we can see data of small triangular.

Quick slicing algorithms create a data structure of triangular with vertices and edges. After the data structure is created, software will reorder all distinct vertexes, and then topology data are completely setup. Finally, the above-mentioned interpolation functions can be used to show the results.

4. Instances of color objects with the process

In this section, two kinds of instances of color objects created with the process are demonstrated, showing the color reproduction and further potential applications.

4.1. UV-ink-based 3D Printing

UV-ink-based 3D Printing follows the similar principle as the DOD (drop on drop) 3D Printing technology. A UV-LED flatbed printer is used as a 3D printer and the UV ink as the modeling material. The UV ink jetting from the printer nozzle solidifies after being irradiated by a UV lamp. Finally, the model is shaped through layer-by-layer printing. However, the difference between UV-ink-based 3D Printing and DOD printing is that the modeling process of each layer of the former is an integrated section and of the latter is a series of lines. So, because of this characteristic of UV-ink-based 3D Printing, it is more suitable for 2.5D modeling. Below are two examples of UV-ink-based 3D Printing.

4.1.1. Application for oil painting reproduction

3D Printing is a novel method to reproduce oil paintings so that it can restore the stereo brushwork of the original. In this work, we use a laser scanner to detect the surface morphology of an oil painting and first build the 3D model. Second, we extract the cross section from the 3D model with a certain contour interval according to the thickness of UV-LED ink. Finally, we print white ink layer by layer for 3D shaping and print a color image onto the neutral 3D model.

We selected an oil painting (**Figure 9**) of 7.5 cm (L) × 6 cm (W) for scanning and printing. A laser 3D scanner is used to obtain point cloud data of the original oil painting (**Figure 10**). Next, we measured the thickness of the UV ink to determine the number of the printing layers.

Figure 9. The original oil painting.

Figure 10. Point cloud model of the oil painting.

This work first conducted the UV-LED printing for 1–10 layers of C, M, Y, K, and W ink separately. Then, we measured the average thickness of the paper and every color lump by a CHY-C2 thickness gauge (see **Table 1**).

Layers	Color lump	0	1	2	3	4	5	6	7	8	9	10
Average thickness (μm)	C	243.9	251.7	261.6	272.6	283.2	294.4	307.1	318.2	328.5	339.2	351.8
	M	243.9	255.7	268.9	282.6	296.2	309.5	322.5	335.4	348.3	361.5	374.6
	Y	243.9	247.7	255.8	264.6	272.1	279.8	287.4	295.9	303.7	310.9	318.6
	K	243.9	248.9	258.0	266.8	276.3	285.1	294.0	303.4	312.1	321.3	330.5
	W	243.9	254.8	268.6	282.9	296.9	310.1	323.8	337.3	350.2	364.1	378.9

Table 1. Average thickness of every color lump.

Obviously, the thickness of these four ink films is different. The average thicknesses of C, M, Y, K, and W ink films are 11, 13.2, 7.9, 9, and 13.7 μm. According to the subtractive color-mixing principle, most radiation light will be absorbed after color mixing. White ink possesses both highest lightness and thickest ink film, so it is considered the best modeling materials for layered printing.

After 3D model slicing, 2D cross-sectional data can be obtained by intersecting the 3D model horizontally with a series of planes. The number of slices that we can obtain in this case is equal to the ratio of model height to white ink-film thickness. As a result, 91 layers are determined to be printed. **Figure 11** shows the result of cross sections extraction. Through contour vectorization and RIP, the neutral 3D model (**Figure 12**) and the reproduction (**Figure 13**) were printed by a UV LED printer.

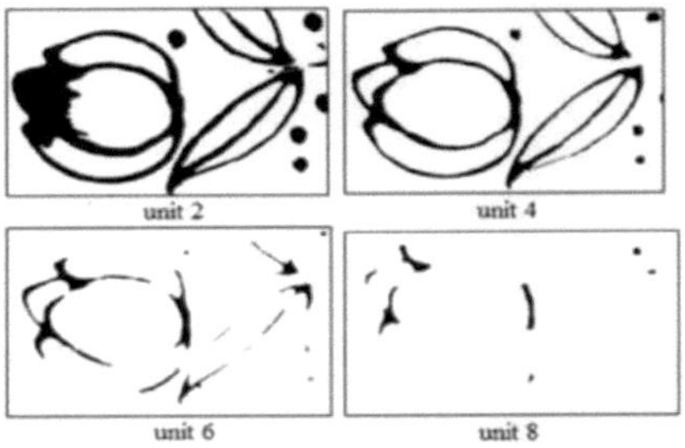

Figure 11. Cross sections.

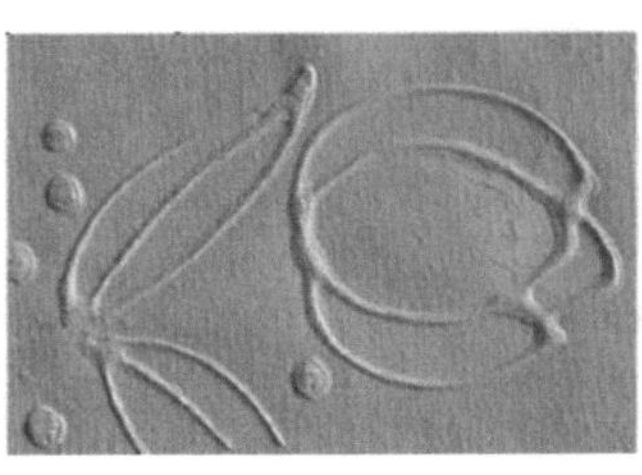

Figure 12. Neutral 3D model.

Figure 13. Reproduction.

4.1.2. Application for a topographical map of Taiwan

3D Printing of a colored topographical map follows the same principle as printing a stereo oil painting. First, we extracted and vectorized the contours of the topographical map of Taiwan. Then, we provided different colors to various heights of contour sections according to the standard of altitude tinting. Based on the actual situation, in order to obtain the best printing effects and quality, 14 layers were tinted as shown in **Table 2**.

Altitude	C/%	M/%	Y/%	K/%	Color
<0	70.0	15.00	0.00	0.00	
0–100	90.0	30.00	95.00	30.00	
100–200	90.0	20.00	100.00	20.00	
200–300	85.0	10.00	100.00	10.00	
300–400	75.0	0.00	100.00	0.00	
400–500	50.0	0.00	100.00	0.00	
500–600	20.0	0.00	100.00	0.00	
600–700	5.00	0.00	90.00	0.00	
700–800	0.00	20.00	85.00	0.00	
800–900	0.00	40.00	85.00	0.00	
900–1000	0.00	55.00	95.00	0.00	
1000–2000	0.00	70.00	90.00	0.00	
2000–3000	0.00	90.00	85.00	5.00	
3000–4000	15.00	100.00	90.00	10.00	

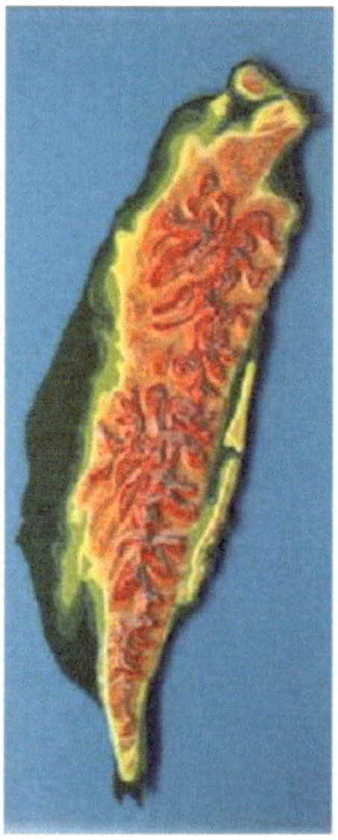

Table 2. Altitude tinting of the topographical map.

Finally, we use UV-LED printer to print the topographic relief and altitude-tint legend. The horizontal scale is 1:3,000,000, the vertical scale is 1:1,500,000, and the printing height is 2.59 mm.

4.2. Paper-based 3D Printing

Matrix 300A, a kind of paper-based 3D printer, is provided by MCOR technologies. It contains a standard Epson 310N printer supplied with the color system and a matrix printer responsible for elevation building. An A4 paper is used and the aqueous adhesives used to adhere two layers volatilize no harmful gas.

Figure 14 shows the flowchart of printing steps. The process begins with a 3D digital model in "*.stl", "*.obj" or "*.wrl" format. Each edge of the layer is printed with bar codes, which are used to read the sequence. Before the 3D building, a piece of a blank paper is affixed to the build plate, and the glue spreader will uniformly coat aqueous adhesive on the paper. Then, the colorful layers are laid on the paper tray of the matrix printer, and the top layer will be pulled by paper-feed roller and gripper onto the built plate. As shown in **Figure 14**, Step 3, the build plate will rise to the top of the build space to press the heat plate, thereby smoothing the top layer and ensuring that it will adhere firmly to the blank paper at a high temperature. The blade will cut the top layer into the contour of the corresponding 2D cross section once the build plate lowers, and the glue spreader will coat aqueous adhesive on the top layer. Similarly, the next layer is processed; the process is repeated until all the necessary layers are processed. When matrix printer stops working, a large paper cube as shown in **Figure 14**, Step 4, is obtained. After the basal and excess portion is removed, the 3D object is obtained.

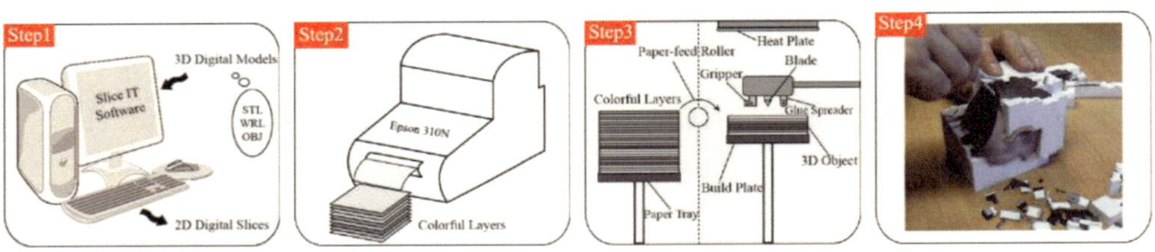

Figure 14. The printing process of the Matrix 300A.

Using paper as a printing material, paper-based 3D Printing shows an excellent performance on full-color reproduction. In fact, the printer only colors the surface of object, the excess and interior portions are all blank, which contribute to low-cost and high efficiency.

4.2.1. Application for historic preservation

Cultural relic is a very important part in the development of human culture. When visitors enter the scenic spots and historical sites, it is often difficult to see the full view of the cultural relics. Because of long time, corrosion, or man-made damage, cultural relics are difficult to save. 3D scanning technology provides a good permanent preservation of cultural relics, and 3D Printing technology provides a good method to produce a copy.

Take a Buddha sculpture for instance, it describes how paper-based 3D Printing technology can do the trick in historic preservation through 3D data acquisition, processing, and printing. The first step is to use the Mephisto EX PRO 3D scanner to obtain the 3D data of a Buddha, as shown in **Figure 15**. The second step is to use the Mephisto software to process the data. It mainly includes removing outliers, smoothing, and data compression, as shown in **Figure 16**. The final step is fabrication. The paper-based 3D printer of Matrix 300A is actuated to finish the reproduction of the sculpture. Basic workflow includes coloring-carving-sizing-pressing-bonding, and the physical 3D model is shown in **Figure 17**.

Figure 15. Scanning process.

Figure 16. Processing photograph.

Figure 17. Physical 3D model.

4.2.2. Application for topographic maps

Digital elevation model (DEM) plays a vital role in engineering construction, hydrology, geological prospecting, etc. Although considering preservation and update of map data, it could not convey actual and accurate 3D feelings to people. In addition, the traditional mold is time-consuming and costly. Thereby, an ability to easily build physical topographic map is necessary. 3D Printing provides the support.

Figure 18. Monochrome physical 3D model.

Figure 19. Colorful physical 3D model.

Slice IT that supports Matrix 300A is used to slice 3D digital model into many 2D layers with equal thickness, the thickness of the A4 paper is a fixed number, 103.2 μm. The 3D solid model of DEM is sliced into 293 layers. After the printing process of Matrix, the physical 3D model is obtained, as shown in **Figures 18** and **19**.

5. Conclusion

A workflow of color 3D Printing is presented here. In view of the current status of color 3D Printing, the theory of color 3D Printing process to print the interpreted data into objects and transfer color information along the way is demonstrated. In addition, specific mechanics from 3D modeling to "ready to print" are described. In order to help readers understand the theories vividly, the research results of our team from South China University of Technology are also listed for reference. The color 3D Printing will become the new engine of innovation driven by Industrial 4.0, and its key links are color rendering and delivery. All of these require cross-supply of unlimited color solution, which is our next objective.

Author details

Guangxue Chen[1,2*], Chen Chen[1], Zhaohui Yu[1,2], Hao Yin[1], Liuxi He[1] and Jiangping Yuan[1]

*Address all correspondence to: chengx@scut.edu.cn

1 State Key Laboratory of Pulp and Paper Engineering, South China University of Technology, Guangdong, China

2 Shenzhen Engineering Laboratory of 3D Printing Technology, Shenzhen, China

References

[1] Fullerton J N, Frodsham G C, Day R M. 3D Printing for the many, not the few. Nature Biotechnology, 2014, 32(11):1086–1087.

[2] Carl S, Langeveld M C V, Donoso L A. Innovations in 3D Printing: a 3D overview from optics to organs. British Journal of Ophthalmology, 2014, 98(2):159–161.

[3] Amado A, Schmid M, Levy G, et al. Advances in SLS powder characterization. Group, 2011.

[4] Shirazia S F S, Mehralib M, Gharehkhania S, et al. A review on powder-based additive manufacturing for tissue engineering: selective laser sintering and inkjet 3D Printing. Science & Technology of Advanced Materials, 2015,16(3):33502-33521.

[5] Daniela N S, Marilia G D O, Eduardo M, et al. Dimensional error in selective laser sintering and 3D-printing of models for craniomaxillary anatomy reconstructions. Journal of Cranio-Maxillo-Facial Surgery, 2008,36(8):443-449.

[6] Dudek P. FDM 3D Printing technology in manufacturing compositeelements. Archives of Metallurgy & Materials, 2013, 58(4):1415–1418.

[7] Shao Z K, Jiang Y L. Key technologies of SLA 3D Printing. Journal of Mechanical & Electrical Engineering, 2015.

[8] Andersson M, Brajnovic I, Pettersson A. System and arrangement for production and insertion of a dental bridge structure, US Patent 8186999 B2. 2012.

[9] Im Y G, Chung S I, Son J H, Jung Y D, Jo J G, Jeong H D. Functional prototype development: inner visible multi-color prototype fabrication process using stereo lithography. Journal of Materials Processing Technology. 2002,130(02):372-377.

[10] http://mcortechnologies.com/mcor-iris-becomes-first-3d-printer-o-embrace-international-colour-standard-for-unprecedented-3d-printing-colour-accuracy/ [Accessed December 2015].

[11] Yuan J P, Chen G X. Speedup method for paper-based 3D color printing based on STL File. Applied Mechanics and Materials. 2015, 731:269–272.

[12] Yuan J, Chen G, Liao J, et al. Visualization of Large-Size Model Based on Paper-Based 3D Printing. Advanced Graphic Communications, Packaging Technology and Materials. Springer, Singapore, 2016.

[13] Chen C, Chen G X, Yu Z H, et al. A new method for reproducing oil paintings based on 3D Printing. Applied Mechanics & Materials, 2014, 644–650:2386–2389.

[14] Wang H M, Chen G X, Zhang W B. 3D Printing of topographic map based on UV inkjet printer. Applied Mechanics & Materials, 2013, 469:309–312.

[15] Xiao K D, Zardawi F, Noort R V, et al. Developing a 3D colour image reproduction system for additive manufacturing of facial prostheses. International Journal of Advanced Manufacturing Technology, 2014, 70(9–12):2043–2049.

[16] Cheung C L, Looi T, Lendvay T S, et al. Use of 3-dimensional printing technology and silicone modeling in surgical simulation: development and face validation in pediatric laparoscopic pyeloplasty. Journal of Surgical Education, 2014, 71(5):762–767.

[17] Yang, J, Wu, L, Liu, J. Rapid prototyping and fabrication method for 3-D food objects, US Patent 6280785. 2001.

[18] Foodjet. Retrieved from http://foodjet.nl/ [Accessed December 2015].

[19] Jia F, Wang X, Mustafee N, et al. Investigating the feasibility of supply chain-centric business models in 3D chocolate printing: a simulation study. Technological Forecasting & Social Change, 2016,102:202-213.

[20] Sun J, Peng Z, Zhou W, et al. A review on 3D Printing for customized food fabrication. Procedia Manufacturing, 2015, 1:308–319.

[21] Stanic M, Lozo B, Svetec D G. Colorimetric properties and stability of 3D prints. Rapid Prototyping Journal, 2012, 18(2):120–128.

[22] Liu-Xi H E, Chen G X. 3D color printing based on UV ink-jet technology. Packaging Engineering, 2015,36(9):134-138.

Chapter 3

Additive Manufacturing of Casting Tools Using Powder-Binder-Jetting Technology

Daniel Günther and Florian Mögele

Additional information is available at the end of the chapter

http://dx.doi.org/10.5772/62532

Abstract

This chapter presents the use of the additive manufacturing (AM) method powder-binder-jetting (PBJ) for the generation of molds and cores for casting applications. Theoretical information on process steps and the binding of particles is given. For the based chemistry, examples are presented and the transfer from conventional production processes to powder-binder-jetting is explained. For sand as a material for metal casting, methods to determine key properties for the casting are in focus. The figures found by our research activities reach up to that of conventional production and indicate the readiness of the technology. These cavities produced by powder-binder-jetting can even be used for cold casting materials such as concrete. Several demonstrators show the impact of this application. A polymer process for the investment casting process is topic of an additional section. To understand this process, theoretical information and figures determined by measurements are presented.

Keywords: Additive manufacturing, powder binder jetting, casting, mold, cavity

1. Introduction

Powder-binder-jetting (PBJ) technology, an important process within the additive manufacturing (AM) technology, may be considered as the standard method for the production of cores and molds for the foundry industry. Despite the enormous turnover volumes, the characteristic variables of the process and the process chain for metal casting are still relatively unknown. This chapter lists the main processes involved in this technology and provides extensive, application-relevant measurement values and results for foundry processes. Thus, the foundry world and the world of additive manufacturing (AM) are closely interlinked.

Additive manufacturing processes differ fundamentally from the conventional production process: The desired component, for example, a mold, is composed of layers. The production is directly related to the component data. In the case of the conventional manufacturing, the component data are at first realized as a model or tool, and this part is then further used for the actual production. The time required for an entire processing cycle is significantly reduced by applying additive manufacturing methods (see [1]).

This layer-based powder-binder-jetting technology produces porous material components by selective bonding of particle material. These characteristics (specifically strength) are disadvantageous for the direct use, but parts manufactured in this manner can be used in the foundry process. A temperature-resistant material, such as sand, is used as particle material for this process. The binder is also selected in accordance to its properties and temperature resistance properties are adjusted. Due to the porosity of the parts, the resulting gas can escape during casting, which eliminates casting defects by rising gas bubbles in the melt.

The particular process associated with powder-binder-jetting which makes this process highly efficient is the use of particle material as a basis and the use of a limited amount of binder. In addition, the automation process can easily be expanded. It is possible to produce component sizes that are specific for foundry technology. Scalability during production allows production rates that are economical for small series. Here as well, a lot of potential can be seen, and previously limited lot sizes increasingly move toward larger lot sizes (see [2]).

This chapter introduces the powder-binder-jetting method in detail and explains its suitability for the production of molds. Other additive manufacturing processes are briefly described and their different characteristics are shown. It is shown that powder-binder-jetting processes based on lot sizes and productivity are unique, and therefore hold a prominent position as an industrial means of production within AM.

The sand casting process plays an important role in the field of conventional casting technology. This sand casting process is carried out using three materials/binding systems, which can be directly transferred to AM production methods. The basic methodology of determining the characterizing parameters are explained further in the following sections. The data obtained for printed components using this method are compared with the data of conventionally manufactured components and an assessment is carried out.

A subchapter deals with the application of 3D printed molds for metal casting in the area of cold-curing materials. Here, the problems arising out of de-coring are most crucial. The process of extracting the printed part from the mold is explained in detail using the examples of concrete casting. The results that are obtained in this area are presented in detail and discussed in the context of applications in structural engineering and architecture.

Another section presents the use of the particle material, PMMA powder, as a basis for the investment casting process. Here, the investment casting process is explained and its embodiment as AM process is explained using 3D-printed forms. Within this context as well, the property requirement necessary for the 3D-printed parts is also elaborated.

2. Powder-binder-jetting

2.1. Overview

This section provides an overview of the powder-binding-jetting technology. As in the case of several additive manufacturing processes, this process too comprises individual steps that are repeated continuously to form a three-dimensional component. **Figure 1** provides an overview.

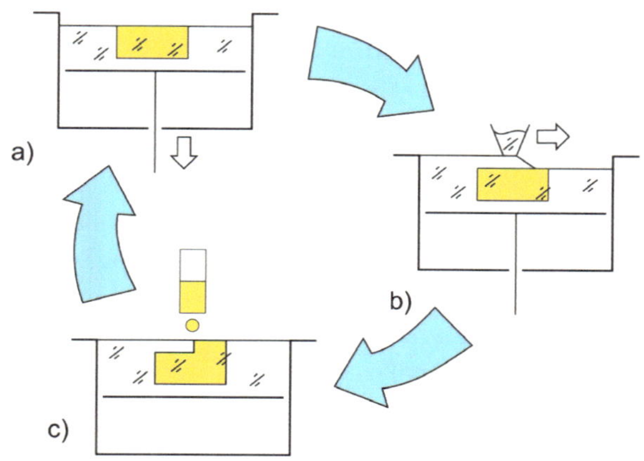

Figure 1. Schematic representation of the powder-binder-jetting-process: (a) lowering the build platform, (b) layering with particle material, and (c) printing process using binder.

The PBJ process steps are as follows: applying a layer of particle material (powder) on a build platform, bonding of the powder particles by a liquid, and lowering the build platform by the height of the desired layer thickness (see [3]).

The process begins with an empty build platform. It is usually set up within a build box. A seal prevents leakage of powder between the moving construction platform and the static walls of the build box. The recoater spreads the powder on the build platform (see [4]). The powder can either be placed in front of the recoater or is moved along with the recoater across the build platform.

At the start of the construction process, the space between the recoater blade and the build platform is sequentially built up with multiple layers of powder to eventually form an even layer of powder.

In the next step, the inkjet print head moves over the powder layer and doses binder onto the layer. The current height of the built component corresponds to a cross-section of the virtual component. The inside part of the cross-section is filled up by print data. Depending on the data processing process, the sliced images are eventually converted into a matrix, in this case in a bitmap structure (for data formats, see [5]).

The inkjet print head also represents a matrix-like arrangement and is guided across the entire built platform, depending on the size of the print head in a meandering motion or in one linear drive across the platform. In the simplest case, this matrix-like arrangement is a line of nozzles which correspond to the driving pattern as determined by the data matrix. During the drive, data change in rapid succession, thus forming the desired image.

The droplets are generated in the inkjet print head itself due to rapid pressure fluctuations and expelled through a micronozzle. Thus, a drop formed in the micronozzle is pressed out and leaves the nozzle as a free-flying independent droplet. Due to this process, certain restrictions are placed on the material range that can be applied (see [6, 7]).

The upper limit for the viscosity is currently approximately 30 mPa s. The surface tension should be substantially below that of water, so less than 50 mN/m. The printable material can include solvents, aqueous solutions, oils, or monomers. The media should not expel particles larger than approximately 1 μm so that no blockages are formed in the microjets.

The final step in the construction process is the lowering of the build platform. After lowering, the resultant space is again filled with the help of the recoater. The layer thickness corresponds to the vertical distance the build platform moved while lowering. An alternative process is the continuous 3D printer. According to this method, the printing is performed on an inclined plane and the powder-bed movement takes place via a conveyor belt at an angle to the plane (see [8]).

2.2. Distinction from other processes

The MultiJet modeling process works similar to the PBJ process. The focus of the method is also an inkjet print head that works on matrix basis. Contrary to the method described above, powder is not used, but the desired shape is composed by the medium expelled from the print head. Here, the medium is cured in layers with the help of a UV source. Overhangs can be achieved by using a support medium that is applied over a second print head. Curing is done using UV technology (see [9]).

Under the vector-based methods, the fused deposition modeling is the only method that is not based on energy beam technology and is thus similar to the PBJ process. As opposed to the two procedures mentioned above, this method does not involve free-flying droplets. The model is constructed using an extruded filament (see [9]).

In the beam-based method, there is no additional mass involved during construction process, reflecting the total mass of the finished model. Existing volume is modified. During stereolithography, resin is cured by a laser beam [10]. When laser sintering process is applied, a powder is used as a base which is similar to the PBJ process and a high-energy beam is used to fuse the powder; Currently, the most widely used methods are the ones making plastic or metal parts. Materials such as ceramics, concrete, or biological materials are the subject of research on an industrial scale but not yet main stream (for ceramic see [11]). **Table 1** provides information on the different distinction possibilities.

Method	FDM	LS	SLA	PBJ
Vector based	X	X	X	
Matrix based				X
Powder based		X		X
Beam based		X	X	
Size	++	0	+	++
Performance	0	+	+	++
Costs	++	0	0	++
Direct application	++	++	+	0
Material diversity	+	+	0	++

Table 1. Characteristics and categorizations of various AM methods.

2.3. Scalability

Among all processes, the powder-binder-jetting process is considered the most scalable. This has an impact on possible component sizes and system performance. Different categories of scalability are discussed and evaluated as below.

In the PBJ process, the build size is defined by the distances covered by the linear axes. In principle, these axes can be extended as required. Merely the axle supporting structure needs to be adjusted according to the increased point of deflection. Limitations are generally not posed by the device, but by the strength of the product produced relative to its weight. This ratio largely determines the handling of the parts and is therefore size restricted.

The performance of a system is often quantified on the basis of the time taken to print one layer.

During the PBJ process, the pixels of a layer are printed by individual parallel controllable nozzles. Here, the number of the nozzles is proportional to the print performance. When measures are taken to parallelize the coating process, the number of nozzles is often nearly proportional to the time per layer. The performance of the printing system can thus be easily scaled by increasing the number of nozzles.

Within PBJ process, there is another option to improve the performance: Several spatially successively arranged coating and printing units are able to generate quasiparallel layers. The applicable layer time is then the layer time of such a "layer unit" divided by the number of the active "layer units" [12].

For procedures having a vector control, such scales cannot be realized. In laser-based methods, the radiation field cannot be expanded indefinitely. This is mainly due to the flat field lens required for a focused exposure. Similarly, it is not possible to use arbitrarily many laser sources simultaneously since the beam path would rapidly become complicated, particularly due to the necessary deviation.

In the case of FDM, axis scaling is possible as which the PBJ processes. However, due to the mass inertia of the print-head discharge, speeds are limited. A parallelization of several print-head discharges is possible albeit only with considerable effort.

The parallelization described by several "layer units" is not possible in the laser sintering process and the SLA process. Both methods use contour procedures that require a complete "visibility" of the current layer. A parallelization would require at least the simultaneous formation of a new layer and the laser exposure of an old layer. The FDM method combines the shaping and layer formation. In this case, this principle can also not be used.

The PBJ method thus provides unique opportunities to improve performance. A summary of these bearings can be found in **Table 2**. Thus, applying this technology as a production process within established production procedures seems to be realizable.

Process	Size	Layer speed	Z-speed
PBJ	++	++	++
FDM	++	0	-
SLS	-	+	+
SLA	0	+	-

Table 2. Assessment of the scalability of different processes.

2.4. Basic modeling

The mechanical properties of the PBJ processes are, apart from the cohesion of the binder and the adhesion of the binder to the particles, essentially determined by two variables: the grain size and the relative amount of binder. During the model forming process, the particles are thought to be stronger than the bond.

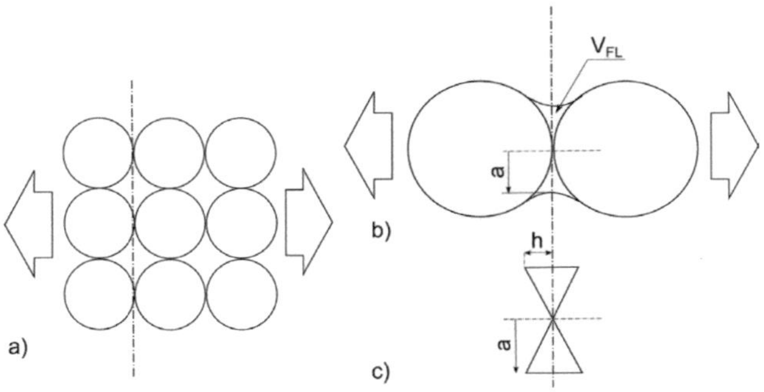

Figure 2. Schematic representation of the model particles/bond: (a) illustration of a particle cluster under tension, (b) illustration of two particles under tension, and (c) highly simplified model for calculating the volume of the bond.

A model of the bond can be represented as in the following simplified diagram: Between two particles that are considered as ideal for the spherical model, there exists an external nearly cylindrical binder bridge (see **Figure 2**). This bridge is formed by capillary action shortly after application of the binder by the print head and the respective strengthening mechanism freezes this condition. This form of connection is similar to the sintering process. Basic deviations, necessary for the basic understanding of the conditions, can be obtained by simple considerations.

If the binder bridge is small in relation to the radius of the sphere (h = const), then considering the above assumptions, a model of a cylinder minus two cones describes

$$V_{Fl} = 2 \cdot \tfrac{2}{3} \cdot a^2 \cdot \pi \cdot h \tag{1}$$

the volume of the bond or the printed quantity of liquid. If the model is modified to determine the influence of the volume, it can be assumed that a constant distance of the particles exists. Only the radius of the cylinder varies. Thus, the relation holds true:

$$V_{Fl} \sim a^2. \tag{2}$$

The contact area A and the maximum mechanical tension σ_{Max} are accessible via simple equations

$$A = a^2 \cdot \pi \text{ and } \sigma_{Max} \sim A. \tag{3}$$

The relationship of the printed amount to the mechanical stress at break is thus linear:

$$\sigma_{Max} \sim V_{Fl}. \tag{4}$$

The strength is thus linearly dependent on the volume of the printed material. The printed volume corresponds to the organic or inorganic binder content minus the evaporated solvent content.

Considering the influence of the particle size, the model must be mentally expanded to form a particle cluster. As a minimal model, a cubic packing may be used and a voxel as observational volume. A voxel is defined by the smallest possible volume that can be produced with a PBJ system. The dimensions of a voxel are made up of layer thickness and the resolution of the inkjet print head together in two directions. For the number of binding areas per voxel N holds true:

$$N = 6 \cdot N_P = 6 \cdot \left(\frac{l_V}{D_P} \right)^3 \tag{5}$$

with the number of particles per voxel N_p, the dimension l_p, a cubic voxel, and the grain diameter D_p of the considered perfect monomodal particle size distribution.

For the volume of a single bond V_B and the total binder volume per voxel V_v, the following equation applies:

$$V_V = V_B \cdot N = V_B \cdot 6 \cdot \left(\frac{l_V}{D_P} \right)^3. \tag{6}$$

As with the approximation of the print volume, in this case only one-half, it is again assumed to form a binding as shown below:

$$V_B = 2/3 \cdot a^2 \cdot \pi \cdot h. \tag{7}$$

For calculating the bond area, relevant for the maximum mechanical stress at break, the following equation applies:

$$A = a^2 \cdot \pi \cdot \left(\frac{l_V}{D_P} \right)^2 \tag{8}$$

and by applying the above relationship for V_B, we obtain

$$A = \frac{3 \cdot V_B}{2 \cdot h} \cdot \left(\frac{l_V}{D_P} \right)^2 = \frac{3}{2 \cdot 6 \cdot h} \cdot V_V \cdot \left(\frac{D_P}{l_V} \right)^3 \cdot \left(\frac{l_V}{D_P} \right)^2 = \frac{V_V}{4 \cdot l_V \cdot h} \cdot D_P. \tag{9}$$

The relationship between σ_{Max} and the particle diameter D_p is therefore, at V_V, l_V, and $h = \text{const}$, almost linear:

$$\sigma_{Max} \sim D_P. \tag{10}$$

The particle diameter is directly proportional to the breaking stress value in the same way as the amount of liquid. This implies that the use of finer particles leads to a reduction of strength. Similarly, when particle size is reduced (improving the surface properties), the amount of liquid printed needs to be adjusted. In the given model assumptions, the density is not affected by changes in the particle diameter.

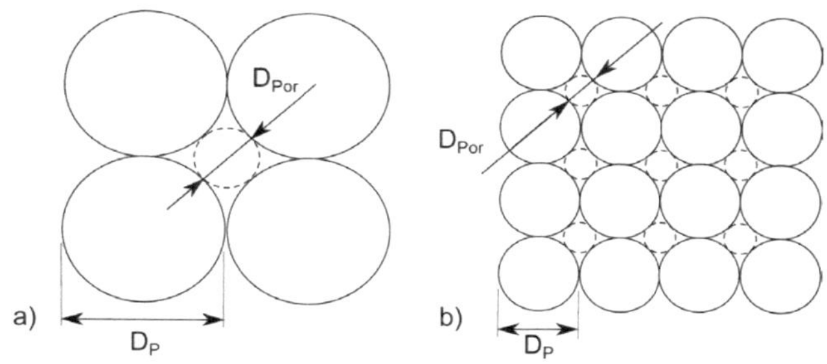

Figure 3. Diagram of the pore-channel diameter at different particle diameters: (a) diameter of an average particle cluster with inscribed minimum passage diameter and (b) particle cluster with half a particle diameter for comparison.

The dimension of the size of the pores is changed by changing the particle diameter (see **Figure 3**). In this example, the diameter of the respective largest circular diameter D_{Por} is inversely proportionate to the particle diameter. This can be roughly described as

$$D_{Por} \sim \frac{1}{D_P}. \tag{11}$$

According to this relationship, the resistance of a gas flowing through such channels greatly increases because the volume flow is proportional to the diameter to the fourth power. A change in the particle diameter (for example to influence surface finish) has a strong influence on the behavior of the gas penetration in the printed model.

3. Production of casts by powder-binder-jetting

3.1. Theory, chemical, and physical basics

The introduction describes the conventional process—to produce metal parts by casting. Sand casting or the combined use of molds and sand cores is explained as follows.

Casting provides important components for industrial production because they provide high functional integration densities with cost-effective implementation. The production of castings is carried out in two steps: First, to create a multipart mold having a cavity which corresponds substantially to the later casting. Second, the casting where the mold is filled with molten metal and after its solidification forms the actual part.

Depending on the complexity, quantity, or cost requirements, the mold can be designed as a disposable or reusable form. Simple structures can be produced using reusable forms. Complex shaped channels in the casting are only possible using disposable molds to realize such complex cores. Often the techniques are hybridized with each other to take advantage of the required assembly (see [13]).

Thus, the cavity of a casting for metallic molds, and molds and cores, may be surrounded by sand. These different parts are mounted before they are cast, some with a large degree of automation (for process steps, see **Figure 4**).

Disposable molds and cores are made from a base molding material that is a particle material using a binder. The binder is often a fluid, which is mixed with the particle material. This mixture is then added to the model and solidifies. The model is shaped in such a way that the model or the core can be removed from the model without being destroyed. Thus, high complexities cannot be achieved using this method. More complex mold sets have to be assembled from individual parts by mounting (see [14]).

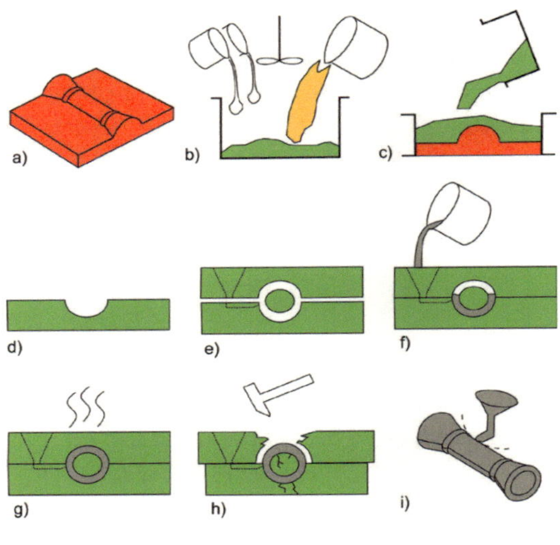

Figure 4. Diagram of the conventional sand casting process: (a) model, (b) preparation of the molding material mixture, (c) molding of the model and solidification of the molding material mixture, (d) part of mold (drag), (e) mounting of molds and cores, (f) pouring liquid metal, (g) solidification and weakening of molding material, (h) extraction, and (i) casting with gating.

For the production of series, the molding material is often mixed with the binder and activated in an automated process (**Figure 4**). This reactive mixture is then injected via a pneumatic conveyer into a model form and gets coated alongside the walls. The component thus produced is cured in the model form. After curing, the part is often automatically removed by a robot and is deburred by an automated process as well. Thereafter, the mold for the actual casting process is again mounted with the help of the robot. The casting process too can thus be carried out fully automated.

The method described requires several basic materials properties for the individual process steps. The particle material must be free flowing to fill the model form. The binder must have very low viscosity for metering and for the resulting molding material flow characteristics (**Figure 5**).

Both properties are also a basis for processing of materials in machines within the PBJ process. Therefore, the materials can be used in 3D printers without major modifications in their physical or chemical properties.

Both methods can also be combined. Forms created with conventional tools or molds are able to support PBJ-printed cores (compare [15] or [16]).

The heat occurring during casting is used to weaken the molds and cores to such an extent that after solidification of the casting the molding material can easily be removed from the casting. This is called de-coring and is significantly defined by the binder.

Among the additive manufacturing methods, the laser-sintering process is the only other AM method by which it is possible to create cores and molds for the foundry industry. In this method, resin-coated sand is melted with a laser and thus selectively bound. The technique is similar to the Croning® method (see [17]).

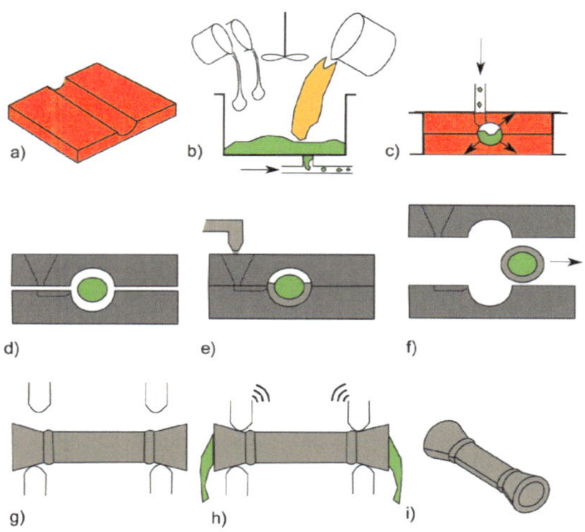

Figure 5. Brief overview of highly automated core production: (a) half section of a core model, (b) preparation of the molding material mixture, (c) core shooting process, (d) inserting the core in the mold, (e) casting process, (f) removal of the casting with the core (gating is not shown), (g) clamping the casting, (h) shake-out of the core, and (i) casting.

Metals that can be used in the sand casting process, in certain cases, exhibit very different properties (see **Table 3**). As far as weight goes, cast iron is the most widely used material for casting. It is characterized by a low melting point of about 1250°C and excellent mold filling properties. Less common casting is the use of steel whose melting point is about 1700°C. Here, form filling is much more difficult. Due to the ready availability of a variety of cast iron types having high ductility, the ductility of the material is only a weak criterion for choosing the appropriate casting material. Therefore, cast steel has become less important. Aluminum is superior due to the lightweight requirements of modern products and has thus become increasingly popular. Aluminum–silicon alloys are excellent for mold filling and melt at around 700°C. In sand casting processes, there are plenty of other metals that can be used. These include the nonferrous metals such as copper, brass, and bronze. Similarly, other light metals or alloys can also be used, for example magnesium (see [18]).

	Aluminum	Cast iron	Steel
Strength	0	+	++
Elongation at break	0	0	++
Damping	0	+	–
Corrosion behavior	+	–	–
Machining	+	+	0
Density	~3	~6.5	~7
Market importance	+	0	–

Table 3. The most important metals in the foundry industry.

3.1.1. Sand and particle material

Particle material is the basis of the powder-binder-jetting process. Therefore, the properties of the particles and the particle clusters are of particular importance and are described below (see **Figure 6**).

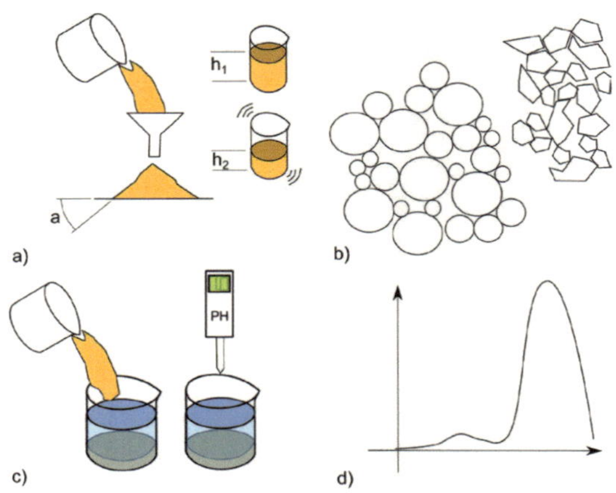

Figure 6. Methods for the characterization of particulate matter: (a) angle of repose and density determination, (b) microscopy, (c) pH value determination, and (d) particle size analysis.

Different groups of sand types are available for sand casting processes. The natural sands are characterized by low cost. The main group forms the quartz sand whose purity permits predictable behavior in the casting. The contents of SiO_2 often exceed 95% in weight. The other

http://dx.doi.org/10.5772/62532

components are usually also oxidic in nature such as Al_2O_3 or MgO. Since the sand prior to use is strongly heated in order to dry, organic admixtures are usually excluded.

For special applications, there are also natural sands which are a mixture of Al_2O_3 and SiO_2. Here, the costs are not as low as with pure quartz sands. Other minerals, for example chrome ore or zircon sands, are common when a high density of the molds or cores is required (see [19]).

Artificial sands, although substantially more expensive, are also common. Here, mostly Al_2O_3 or SiO_2 are used as the base. These sands are chemically highly pure and the particle spectrum is precisely adapted to the requirement (see [20]).

In the case of natural sands, the grain structure depends on the origin of the sands. Cracked products are made up of sharp-edged grains that are generally more square shaped than spherical in shape. Sand qualities can be obtained from natural deposits such as rivers, whose particle shape is nearly spherical. These sands also often exhibit particularly favorable microtopology for binding reactions (see [20] and **Figure 7**).

Depending on the production process, the form of artificial sand is usually spherical. This property allows these products to flow very well and there are smooth surfaces.

Figure 7. Various mold raw materials: (a) SEM picture of natural sand and (b) optical microscopy artificial molding material.

The sand particles react with various chemicals. Many oxides react alkaline in an aqueous environment. Therefore, the pH value of the applied media changes, which is especially significant for PBJ processes when the sand is premixed with a liquid.

A measurement of the chemical properties can be carried out by measuring the pH of the wash solution for the sand. For example, for an amount of water of 100 g, 10 g sand is given and after a waiting period of 10 min the pH value is determined with a pH meter with a glass electrode or is determined titrimetrically against an alkaline solution.

Natural sands have a significantly broad particle size distribution. Here, often a noticeable amount of fine particle distribution of up to <10 μm particle size can be noticed. This affects various properties, among other things, the flow property, but also the bond because the micrograins attract the binder in a capillary manner and thus withdraws from the binding region of the mold-forming grains (find different qualities in **Table 4**).

Undesirably large grains are often separated by sieving. This leads to a sharp divide in the grain size range. The bulk density of the sieved as well as unsifted materials falls way behind theoretical possibilities. With natural sand, about 50% volume space filling can be achieved. Targeted blends can positively affect the bulk density (see [21]).

Artificial sands are nearly monodisperse, i.e., they have essentially only one grain size. This property and the round grain often lead to an increased bulk density. Again, this effect can be enhanced by adding further monodisperse powders having different grain size.

Natural sands are often preferred because of their low cost. However, especially with quartz sands, a change in the crystal structure of the elementary cell (quartz inversion) takes place at a temperature of 573°C. These stretches abruptly produce a sudden increase in volume of the core or mold. This often leads to flaking of layers close to the mold cavity, thus causing casting defects (see [14]).

For artificial sands, this effect is usually only minimal. Here, additionally a high density is often implemented in order to reduce the buoyancy of cores in the melt.

In the PBJ process, much finer sands are used compared with sands used in the standard foundry. This is necessary in order to achieve surface qualities that are comparable to the conventional process because the grains are not aligned toward the surface of the component as is the case while using a model.

	Quartz sand GS14RP	Cera beads AFS100	Kerphalite HA
Chemical base	SiO_2	Al_2O_3/SiO_2	Al_2O_3/SiO_2
Purity/contents (%)	>98	62/38	60/40
Medium grain size (μm)	140	150	200
Specific surface (cm^2/g)	176	172	105
Sinter temperature (°C)	>1550	>1660	>1660
Residual water content (%)	0.2	0.7	0.7
pH of washing solution	6.3	6.7	6.4

* the shown figures are simplified and do not take account of the small amount of additional oxides.

Table 4. Examples of data for sand systems which are suitable for the PBJ process.

3.1.2. Binder systems

The bonding strength between particles is caused by the binder. In the case of PBJ, the binder is selectively applied with the help of an inkjet print head.

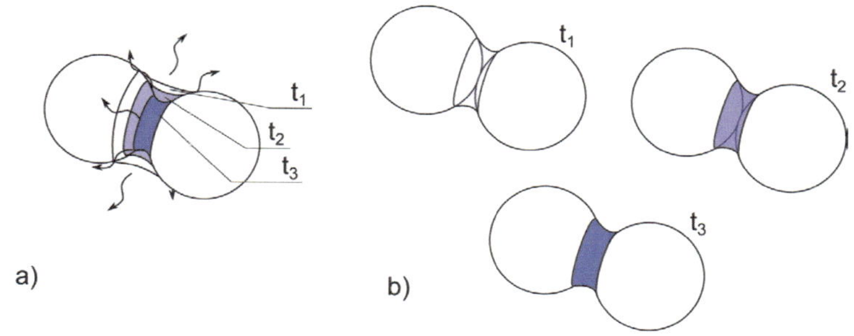

Figure 8. Basic binding mechanisms: (a) solidification by evaporation of solvent and (b) curing by polymerization.

Basically, two binding mechanisms can be used for the above-mentioned PBJ process (as depicted in **Figure 8**).

The material may be dissolved in liquid binder. The adhesive effect of the material is obtained by drying. Often plastics or inorganic binders are dissolved in water or suitable solvents. The drying must take place as long as the layer is exposed and the parts lie covered in powder during the construction process. A variation on this procedure is to dissolve the particle material with the help of a solvent which is part of the printing liquid and then to let it dry again. Thus, a binder-free medium can be printed. This is favorable for the life of the print head. But the problem is often the long setting time.

Polymerizing binders can also be used. Polymerization is a chemical reaction where mostly low molecular weight substances react in longer chains of molecules. Thus, a transition between low viscosities, i.e., good printability and extremely high viscosities (solid), is made possible.

Usually, it is not only necessary but also useful to use both methods as a hybrid. This allows moderate reaction rates that safeguard all the machine components and support a drying process deep inside the large powder cake.

The properties of the binder which are suitable for the PBJ methods are very similar to those of the conventional molding production. In the following, the most common systems are described according to their chemical and physical characteristics (**Table 5**).

Furan resin is a polymer binder which is derived from corn cobs. The resin system is available in the form of a resin and an activator component. The resin itself is made up of monomers and oligomers, for example, furfuryl alcohol, bisphenol A, and resorcinol. In addition, adhesion promoters are included. The activator is an acid, usually a mixture of several acids. It has a catalytic effect on the reaction, but is itself also consumed in the process. The polymer with is produced through polycondensation is a cross-linked, high-strength thermoset which is highly temperature resistant due to furan groups in the molecule (see [14]).

In contrast to furan resins, phenol resins consist of phenol units, which are preferably linked to each other in *para* and/or *ortho*-positions via methylene groups. No continuous conjugated system is formed. The curing process is carried out using formaldehyde depending on the prepolymer. For example this can be made available for the process by the thermal decomposition of the substance Urotropin. A thermoset plastic is again formed. The temperature resistance is even higher than that of furan resin (see [14]).

Inorganic binders are gaining importance not only because of their historical importance, but also because of new requirements for environmental-friendly production. They are characterized by very low and harmless cast emissions. For this process, liquid water glass is of particular importance. It can achieve high strengths. However, the de-coring process after casting requires more effort with this binder system than with the organically bound varieties (see [14]).

	Furanic resin	**Phenolic resin**	**Inorganics**
Bending strength (N/cm^2)	250*	350*	300*
Temp. stability	Sufficient	Very stable	Extremely stable
De-coring	++	+	–
Shelf life (months)	12	6	Unlimited
Emissions	High	High	No

Table 5. Relevant binder systems and properties, *typical figures, depending on the amount of binder.

3.2. Methods for the characterization of PBJ generated parts for foundry purposes

Using PBJ printed components as tools for casting processes requires a special range of characteristics for the parts. The porosity of the parts is essential for the process. Similarly, the strength values are important to ensure safe handling, as well as safe and easy de-coring after casting. In this process, some values are linked and, under certain circumstances, represent contradictory optimization criteria (**Figure 9**).

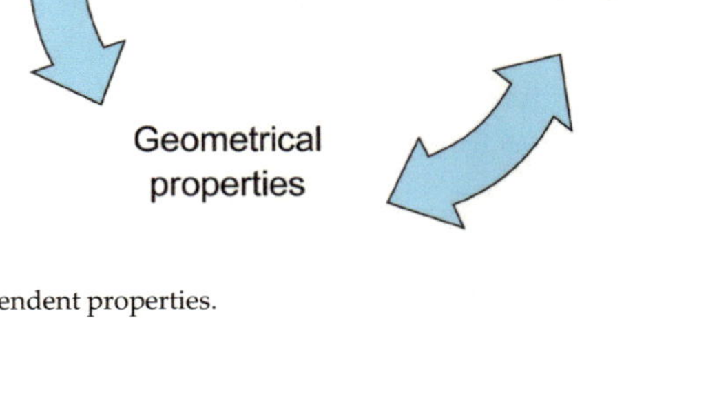

Figure 9. Schema of interdependent properties.

The resulting amount of gas emitted from the burning binder during casting must be diverted through the porous structure of the molds and cores. The volume of the gas impact will depend on the proportion of the binder. This in turn significantly affects the strength. The strength is dependent upon the particle size, the related resolution, and the achievable surface quality of the process in relation to the geometry. Similarly, high strength often means the tendency for shrinkage and thus distortion. The geometry works once again due to the effect of the grain size on gas permeability, as the finer particles cause the gas channels to become narrow.

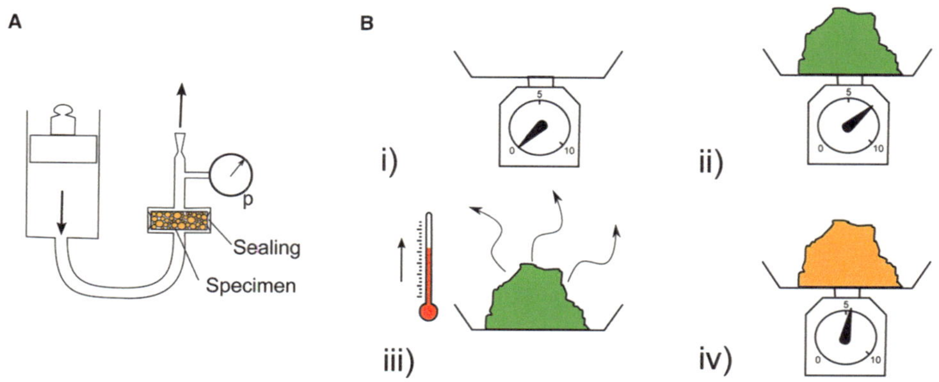

Figure 10. (a) Diagram showing the arrangement to determine gas permeability and (b) process cycle during the determination of ignition loss: (i) taring process, (ii) (initial) weighing material, (iii) glowing, and (iv) weighing material.

The range of printed applications as well as conventionally manufactured cores and molds is affected by gas permeability. To determine the behavior of the gas, the flow rate can be measured with the help of a test specimen at a given pressure. In this case, a tester of the type SPDU of + GF + is used. A cylinder having dimensions 50 mm × 50 mm is used as test specimen (for schematic, see **Figure 10a**).

The second important parameter characterizing the moldings with respect to the gas problem is the loss on ignition. This is determined by taking a sample from a bound form or a core. A furnace of type KLS 05/11 from Thermconcept and scales of the type of PCB2500-2 core were used in the above-described experiment.

According to the standard method for determining the loss on ignition, about 30 g pulverized molding, usually from test bars, is weighed and put into a ceramic bowl. Before the bowl is filled, it is constantly heated to constant weight in an oven to extract moisture and organic residues. Before and after filling, the bowl is weighed. The binder or moisture in the components is burned out or expelled at about 900°C for approximately 30 minutes. After cooling, the weight is determined again. The difference between the weight of sand sample before and after heating is the absolute loss on ignition which is usually expressed in percent by weight on the initial weight (see **Figure 10b**).

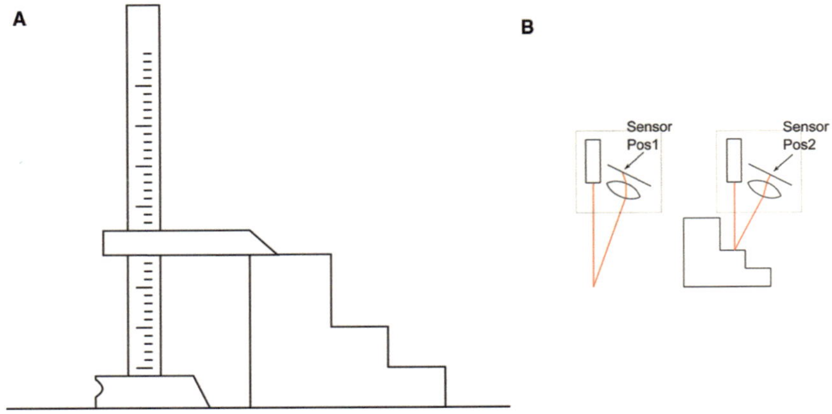

Figure 11. (a) Detecting the main dimensions with a vernier height gauge and (b) basic determination of triangulation.

The accuracy of the construction method is determined based on test specimens. These determine the scope of the machine. Rods measuring up to 500 mm can be measured in this manner. The simplest and most common method is the direct measuring with a caliper or vernier height gauge. More complex shapes can be determined using optical methods. Some of them are based on the principle of triangulation. Here, a device of the type MICRO-OPTRONIC optoNCDT is used (ref. overview [22]) (see **Figure 11**).

The surface parameters of the molds and cores are determined by a tactile measuring method. Here, the extreme microstructure and the weak binding of the grains are taken into account so that an imprint of the surface is made before measurements are carried out (**Figure 12a**).

For the purpose of forming an imprint, plasticine may be used as a quick means yielding good accuracy. Plasticine is soft and produces a clean and accurate imprint. The impression thus obtained is cured at 100°C in the oven. The imprint is then scanned with a surface measuring device, i.e., type Mahr MarSurf SD 26. The evaluation of the values Ra and Rz is calculated automatically (**Figure 12b**).

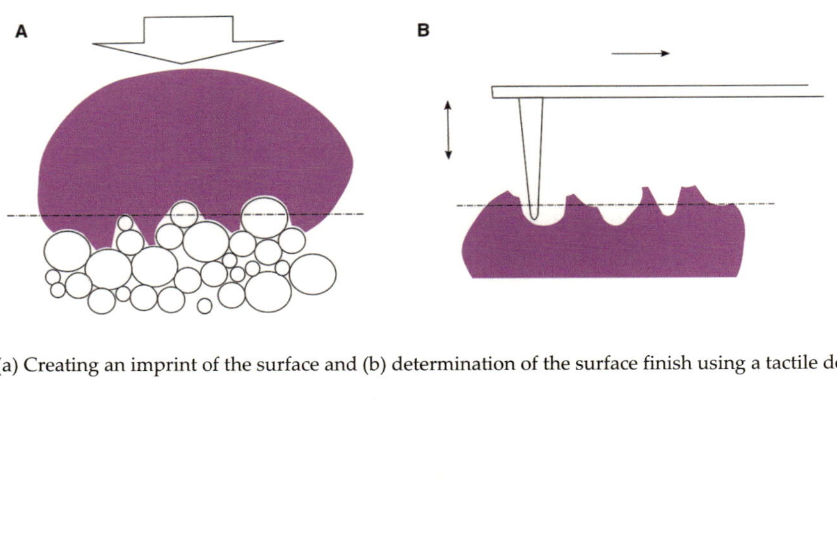

Figure 12. (a) Creating an imprint of the surface and (b) determination of the surface finish using a tactile device.

The strength of 3D printed parts is determined using destructive testing just as in conventional production methods. Conventional test equipment is used. A universal type ZMART.PRO of the company Zwick is used for the determination of the *E* modulus, the elongation at break, and the tensile strength. For a three-point bend test which is commonly used in the foundry industry, a test device of the type +GF+ PFO is used.

A test job is specifically built having the appropriate material system for recording the measured values. This includes all relevant tests specimens in different variations for the described tests. The material system furan is processed on a Prometal S15 system. The phenolic resin and inorganic binding material are processed on a voxeljet VX1000. **Figure 13** represents a test job on a VX1000.

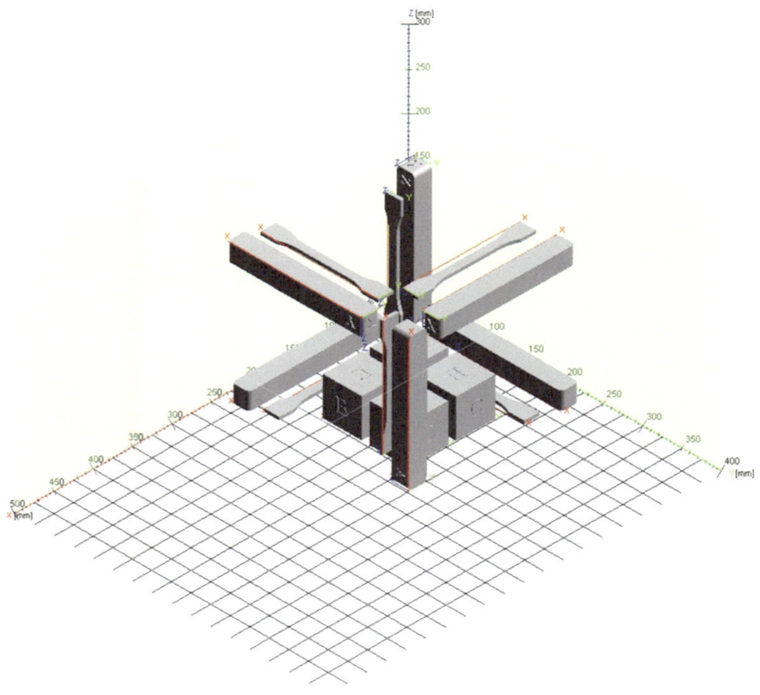

Figure 13. Arrangement of different test samples within a test job.

3.3. Results for PBJ printed parts

Within PBJ, the furan resin system is the most common and long-term proven system. The measured values refer to a system with the sand type GS 14 RP Strobel quartz sand. As sand binder, a binder from ASK type ASKURAN with the corresponding hardener is used.

The test samples show a strength with an average of 270 N/cm^2 at a density of 1.35 g/cm^2. The scattering of 20% within an orientation can be measured. The maximum anisotropy of the strength values is shown in a deviation of nearly 40% in strength for bending-test samples built in the *Z*-direction (see **Table 6**).

Orientation	X			Y			Z		
	Min.	Mean	Max.	Min.	Mean	Max.	Min.	Mean	Max.
Density (g/cm³)	1.33	1.37	1.41	1.33	1.36	1.41	1.32	1.33	1.36
Ultimate bending strength (N/cm²)	222	276	326	236	278	331	219	245	268
Modulus of elasticity (GPa)	1.4	1.3	1.1	1.1	1.3	1.5	1.0	1.1	1.2
Ultimate tensile strength (MPa)	0.6	1.1	1.3	0.8	0.95	1.2	0.2	0.4	0.6
Elongation at break (%)	0.05	0.06	0.08	0.04	0.07	0.09	0.02	0.06	0.16

Table 6. Mechanical properties of PBJ-printed parts.

The measurements of the length of the bars exhibit low scatterings in the range of about 0.2%. The surface quality of differently arranged surfaces widely varies from each other. In this case, the surface roughness Ra of the top surface with 19.8 μm is the lowest. Subsequently, parallel to the *XY* plane, surfaces having a Ra of 22.4 μm follow. The area with the roughest surface is oriented in the *Z*-direction (Ra = 25.2) (**Table 7**).

Orientation	X			Y			Z		
	Min.	Mean	Max.	Min.	Mean	Max.	Min.	Mean	Max.
Surface quality Ra (μm)	19.8	22.0	25.7	21.4	24.3	25.6	22.4	25.2	29.1

Table 7. Geometric characteristics of PBJ-printed parts.

In the settings used for the test job, a loss on ignition of 1.8% can be measured. The gas permeability of the material system is 68 l/h. **Table 8** provides the gas permeability of more sand systems. It is assumed that each system has a for the strength reasonable compaction, and thus also a reasonable density.

	GS14	GS19	GS25
Grain size (μm)	140	190	250
Gas permeability (l/h)	65/75	140	250

Table 8. Gas permeability of PBJ-printed parts using different grain sizes.

A comparative measurement of different chemical binder systems is performed on the basis of a variety of systems, and in each case an optimized configuration for the intended use is applicable. **Table 9** provides a brief comparison of the results.

	Furanic resin	Phenolic resin	Inorganics
Density (g/cm^3)	1.29–1.31	1.29–1.31	1.2–1.25
Sand grain size (μm)	140	140	170
Loss on ignition (%)	1.6–1.9	2.2–2.6	0.2–0.3
Bending strength (N/cm^2)	220–330	350–450	250–350
Surface quality (μm)	24	28.5	23
Length (mm)	172	172	172

Table 9. Properties of PBJ-printed parts using different binder systems.

3.4. Discussion

The special characteristics of the printed components relative to the conventionally manufactured components are the smaller grain size, the alignment of the grains, and built-layer characteristics unique to each method. Thus, the 3D printing itself exhibits different values. At a comparable grain size, the surface finish of 3D printed parts is substantially lower compared with that of the conventional parts. Similarly, the loss on ignition is higher at the same strength, as the conventional methods have a greater densing effect on the material.

The methodology for measuring the properties can be well transferred to the PBJ process. But there are special considerations to be observed. Most dimensions are anisotropic. Additionally, the artifacts "stair steps" have to be considered for an objective surface quality comparison and each surface orientation has a different surface finish.

The use of 3D printed cores is by now widely spread. Even complex cores can be produced by this method with a suitable choice of the binder system, although some characteristics of the conventional production may not be achievable. To produce challenging cores, it is possible to find good compromises for test series. The material systems described above provide such processes, material and binder combinations.

4. Forms and cores for cold setting materials

In addition to metal casting, several important casting processes exist for the production of molds. For example, thermoplastic resins are injected into molds under high pressure. Commodities thermosets usually have very low viscosity and can be poured just like liquid metals. Casting of concrete is also relevant. Here, large volumes and tonnage can be cast. These molds can be built as forms and also be constructed using PBJ-printing technology (see [23]).

In concrete casting, carpenters are often involved in preparing the formwork. The castings are mostly simple geometries. Pillars, walls, and ceilings are predominately made by this casting process. But more complex forms such as curved stairways are poured into wooden molds (see **Figure 14** and compare [24]).

In prototyping, even large components are cast with reactive resins. Here as well, large molds are necessary to give the casting the correct geometry. The casting process is essentially analogous to the conventional concrete pouring.

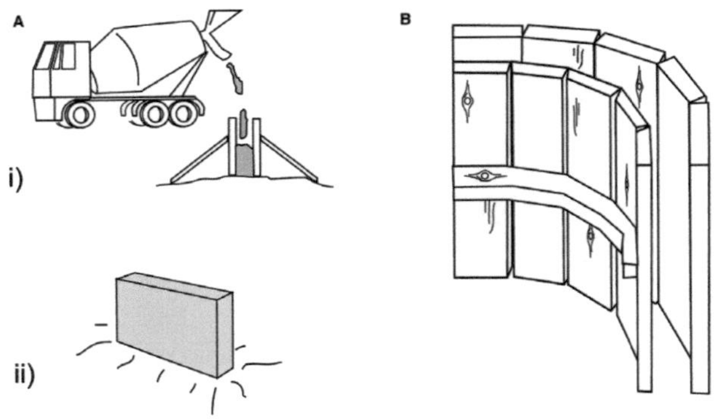

Figure 14. (a) Conventional procedure in construction: (i) pouring concrete into formwork, (ii) finished de-cored wall and (b) construction made by a carpenter: curved wall structure.

Concrete casting usually consist of smooth walls, which can be used repeatedly, and hand-crafted wooden construction, which is often used only for one part. After the mold is set up, the reinforcement is usually built from iron. Then, the previously mixed concrete is poured. After curing, the formwork is cut off and the concrete can be surface treated.

In concrete construction high complexity is seldom realized due to the formwork technology. In contrast to metal casting, the heat of the cast material cannot be used for de-molding.

A high degree of complexity can be achieved with printed forms. Molds may be used once or several times depending on their geometries, for example, whether forms have undercuts or not is a relevant criterion.

The printed forms have to be surface treated irrespective of the base material used (**Figure 15**). There are two essential points that need to be considered. Firstly, it is important to prevent cement from entering the pore space of the mold. This would also lead to the mold sticking to the form and to cement leaking around the peripheral regions. Both these factors mean that the outer surface of the casting will be damaged.

Secondly, the separation of casting and form must be made easier if the form is to be used several times. In this case, the adhesion of various materials to each other is to be considered as well as the exact arrangement as lack of air circulation results in high adhesive forces during the separation process.

Hydrophobic substances such as fats are usually used as release agents. These can be applied directly onto the forms. A free-flowing greasy substance thereby allows through-flow of air from the pore space and enables easy removal from the mold.

In case strong forces are created during unmolding, the forms can be strengthened prior to use by infiltration. For example, this can be done by epoxy.

In some cases, the integration of ejectors into the mold is advisable. These can be implemented as jacking screws. Targeted stresses can be built up in the area of the form in order to reduced bending stresses while separating the mold.

Figure 15. (a) Printed form mounted (form for casting upside down) and (b) designer sink made from concrete.

The mechanical properties are relevant for multiple phases in the application of these processes. Strength is desirable during the handling phase and while filling the mold. While demolding of disposable forms, low strengths are desired. Particularly suitable are shapes and forms whose strength can be influenced after casting. These can be water-soluble forms with a hydrophobic coating or forms whose binder loses strength even at very low temperatures.

The thermal behavior of the forms plays a subordinate role in this process. Similarly, loss on ignition or residual ash is not relevant.

In this process, accuracy and surface quality are just as relevant as in the metal casting. However, the proportions of the surface features of the component dimensions are often so extreme that the surface finish plays only a subordinate role. The achievable accuracies usually do not greatly exceed the general requirements in the construction industry.

See **Figure 16** for an experimental casting of a large-scale structure.

Figure 16. (a) Composite mold from 3D printed formwork elements and (b) resulting large-scale concrete member.

5. Investment casting with positive molds using PBJ printed material

5.1. Theory, chemical, and physical basics

5.1.1. Conventional investment casting

The investment casting process is used to produce metal parts that have finer details, thinner wall thicknesses, and superior surface quality compared with parts generated by the sand casting process. Since the foundation of the process is a positive model, the process differs in its substantial process steps from the sand casting process. See **Figure 17** for details. In this section, various methods are described to create cavities which can then be applied in the metal casting process (compare [18]).

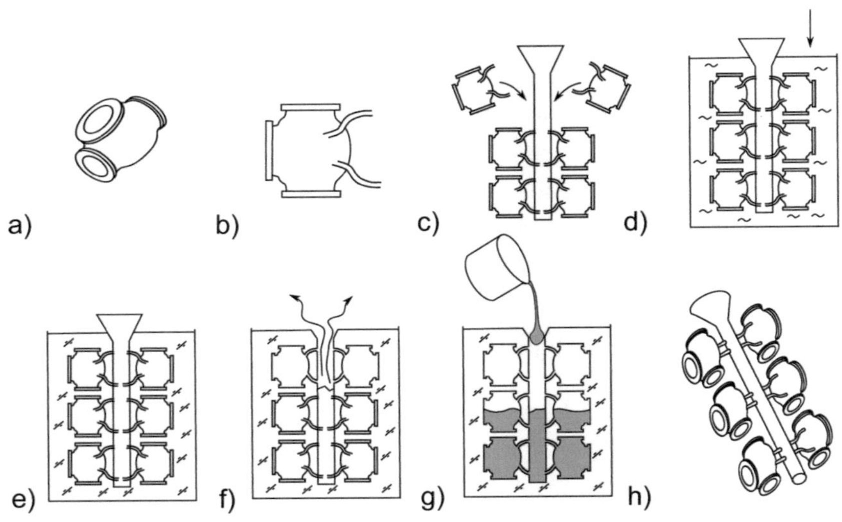

Figure 17. Steps of investment casting: (a) positive model, (b) fixed gating system, (c) mounting of the (wax) cluster, (d) embedding in plaster or ceramics slurry, (e) setting of the shell, (f) burn-out of the model and the gating system with hardening of the shell, (g) casting, and (h) production result.

The basis of the investment casting process is a positive model, which are always lost models in contrast to the models used in sand casting. These models can be produced by artisans manually using the traditional way or by applying industrial mass-production technology.

The models are often produced by injection molding out of wax. As in the plastics processing, the wax is injected under high pressure into a metal mold and removed from the mold after cooling. In order to increase the dimensional stability and to change other properties, the wax is often filled with microgranules. The granules themselves are mostly plastics such as polystyrene or polyethylene.

The individual wax models are attached to further wax model in the production cycle. This wax model serves as a carrier and defines the gating in the subsequent casting of the liquid metal. The individual models are connected using thermal wax to the gating. The upper part of the gating widens conically and later represents the pouring funnel.

The shape thus obtained is further processed in the conventional method in two ways. The first method is to immerse the mold in liquid plaster. After the plaster has hardened, the mold is placed in an oven for heat treatment. Here, the wax evaporates or burns out and exposes the actual casting cavity. The second option is to generate a ceramic shell by repeated applying a coat of ceramic slurry. It is then fired in a kiln. During this process, the model evaporates or gets burnt out.

Both methods differ in the cooling rates which are attained later in the casting process. Especially, the simpler method using gypsum often produces an inferior metal structure and thus a casting of poor strength.

5.1.2. A PBJ process for the generation positive burn-out models

In principle, several methods are suitable for the realization of the above-described investment casting process using additives manufacturing methods. These methods are wax or acrylate direct print processes, laser sintering of polystyrene, or stereolithography. Each method demonstrates specific characteristics which in turn make them the preferred method to apply depending on the production environment and related variables (see [25]). Different levels of components can be built by these technologies: from the single part model up to a complete cluster (see **Figure 18**).

An essential feature underlying all methods is the melt-out or burn-out property of the additive material that is being applied. Here, residue-free burn-out during the process (as far as possible) is important from an economic point of view as purification steps can thus be avoided. Individual melting is not sufficient, as the melt cannot run out from complex shapes and the material is able to leave the channels only in a gaseous state.

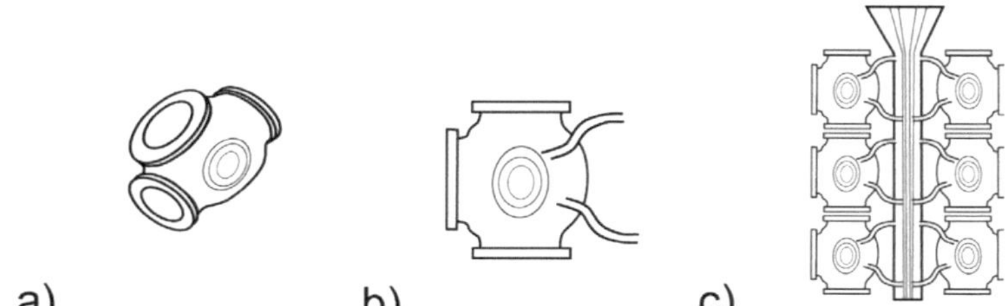

Figure 18. Production levels of components using AM processes for precision casting tools: (a) positive-model, (b) model with sprues, and (c) complete cluster with gating.

The PBJ method uses an acrylate-based printing process for investment casting. PMMA powder serves usually as base material.

The process is described in detail in the subchapter "sand casting." After lowering the build platform, a layer of PMMA powder is spread over the built platform with the help of the recoater. The achievable minimum layer thickness is well below that of the sand casting process and reaches 80 μm. Through a binder system technology, the powder particles stick to each other whereby binder is applied on to the powder bed by an inkjet print head by using preprogrammed bitmap data. For this purpose, solvent-based adhesives can be used as well as polymerized adhesive. For the latter process, one component of a multicomponent adhesive system is added to the powder.

The basic chemical reaction of the polymerizing system is a free-radical polymerization. An initiator, for example, dibenzoylperoxide, is present in the powder itself. The concentration can be adjusted as required during powder production. At higher temperatures, the initiator decomposes into two radicals depending on the type of initiator used.

The actual ink contains monomers. These monomers have radically polymerizable chemical double bonds, which are the key reactants for the chain reaction in curing of such materials (**Figure 19**). In addition, the binder includes an activator which brings about the decomposition of the initiator and at the same time generates initiating radicals. In this case, the activator reduces the temperature of the disintegration under the room temperature. When ink penetrates into the powder, initiator from the powder is released and is activated by the activator. The radicals are bound to the monomers (chain initiation), whereby a new radical is formed, which reacts with another monomer (chain continued) in turn form further radicals. Thus, a chain reaction that leads to macromolecules is formed which makes up the strength of the bond (see micrographs of the bonds in **Figure 20**). The reaction stops (chain termination) if, for example, radicals react with each other (recombination) or the reaction mixture of reactive monomers increasingly reduces leading to a failure in reaction. The achievable strength depends on many factors, which in turn affect the chain reaction (see [26]).

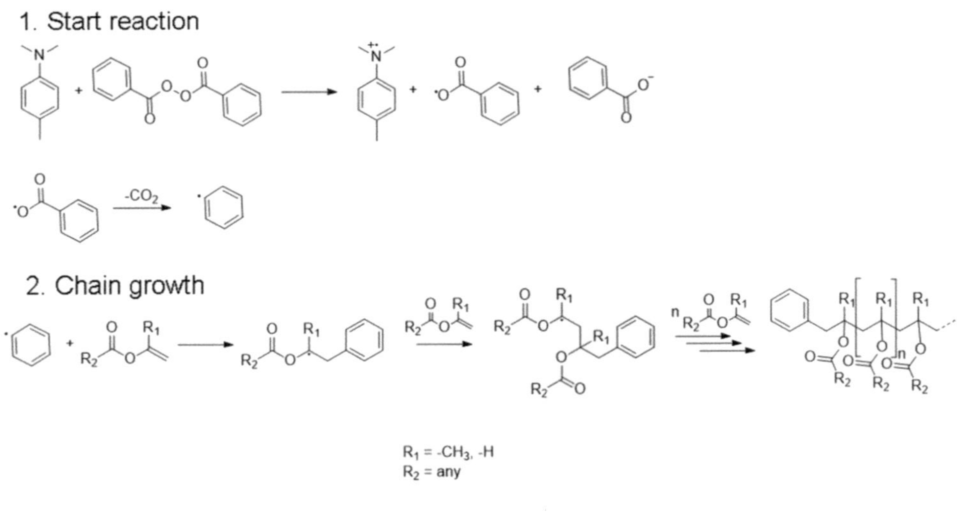

Figure 19. Chemistry involved in the vx-acrylate process.

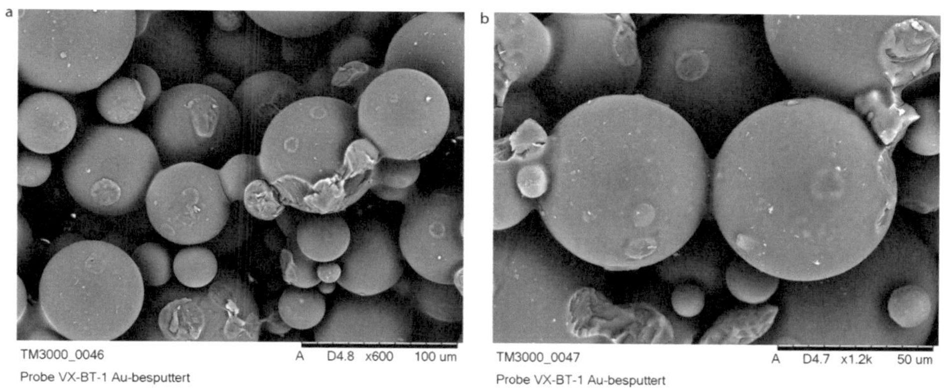

Figure 20. SEM micrographs of particles bound by the PolyPor method (a) magnified 600 times and (b) magnified 1200 times.

5.2. Methods for the characterization of PBJ generated parts for investment casting

In investment casting processes, just as in the case of sand casting, special component properties need to be achieved in order to support further processing steps.

Material strength values are important in order to allow safe handling of the component. But too high material strength values and a high modulus of elasticity are in fact damaging and thus harmful for follow-up processes. Here as well, a low thermal expansion coefficient is important. The mechanical characteristics are determined with a tensile testing machine of the type ZMART.PRO from the company Zwick. Here, the tensile testing program is carried out at room temperature at a speed of 1 mm/s. The measurement is carried out without displacement transducer. As a sample, a tensile testing rod is used with the cross-sectional dimensions 10 × 4 mm^2 (see **Figure 21**). The load cell has a measuring range of 2 kN.

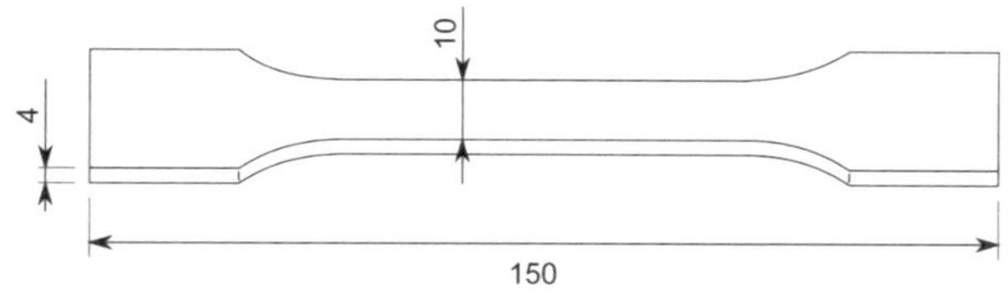

Figure 21. Test specimen dimensions for PBJ-parts used in investment casting.

For the embedding process, a closed surface is required. Without this closed surface, water can penetrate into the model. This could negatively affect the mold during burn-out in terms of cracks or disintegration of the shell. The surfaces of the PBJ-printed models are coated as porous surface adversely affect the follow-up processes of the above-mentioned processes. Thus, it is advisable to treat the model with liquid wax. The wax solidifies on the surface or penetrates by capillary action deep into the model, depending on the immersion conditions and temperatures of the wax or the component. A quick immersion of a cool component leads to a wax layer forming on the surface of the actual component. On the other hand, if a component is immersed over a longer period it gets almost completely filled with wax and has a significantly rougher surface.

The surface must be sealed for the follow-up process. If only a thin surface layer is obtained by the dipping method, there is a risk of microporosities causing surface damage during solidification and shrinkage of the wax (see **Figure 22**). This may lead to water seeping into the model. Thus, the model may be damaged during burn-out. The method of immersing components for a longer time poses the risk of component distortion due to the temperature reaction as well as the creation of rough surfaces.

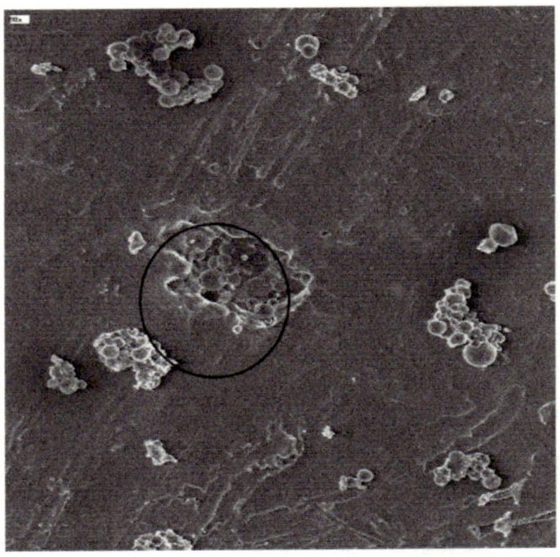

Figure 22. SEM microscopy of a wax treated surface with pinhole defect.

This microporosity can be detected with the dye-penetration technique in accordance with DIN/ISO EN 571-1. Here, the dye accumulates in the microporosity and reveals the critical areas. Such critical areas must be reworked before the furnace process (**Figure 23**).

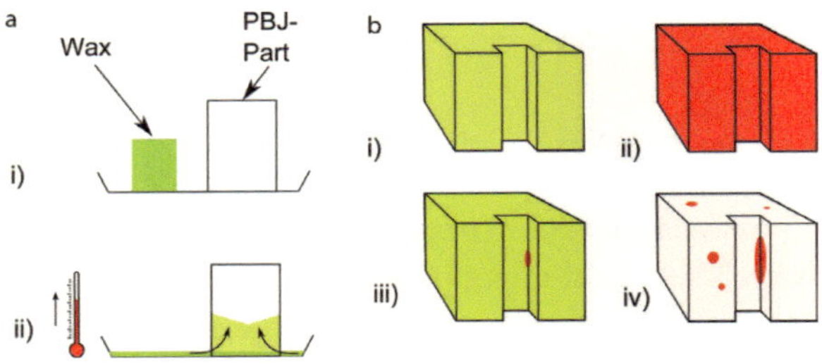

Figure 23. (a) Diagram showing after-treatment of porous components and (b) test: (i) waxed component, (ii) test color, (iii) the component after cleaning, (iv) the component after application of the developer.

After burning out, the residual ash is also an important factor for safe processing. In particular, undertaking cleaning steps after the casting cavity has been formed through burn-out is not advisable economically.

The residual ash is determined using the conventional ash measuring equipment. Since the quantities are very small, the procedure has to be performed with great care (see the steps in **Figure 24**). For the measurement at least 100 g bound material is required. These are added to a previously extensively heated bowl. After weighing, the plate is placed in an oven and heated up to 700°C. This temperature is maintained for at least 4 h. After cooling down for several hours, the dish is weighed again. The difference to the tare value will be read from the amount weighed. The quotient is thus residual ash. A residual ash content of less than 0.1% is considered as free of residual ash.

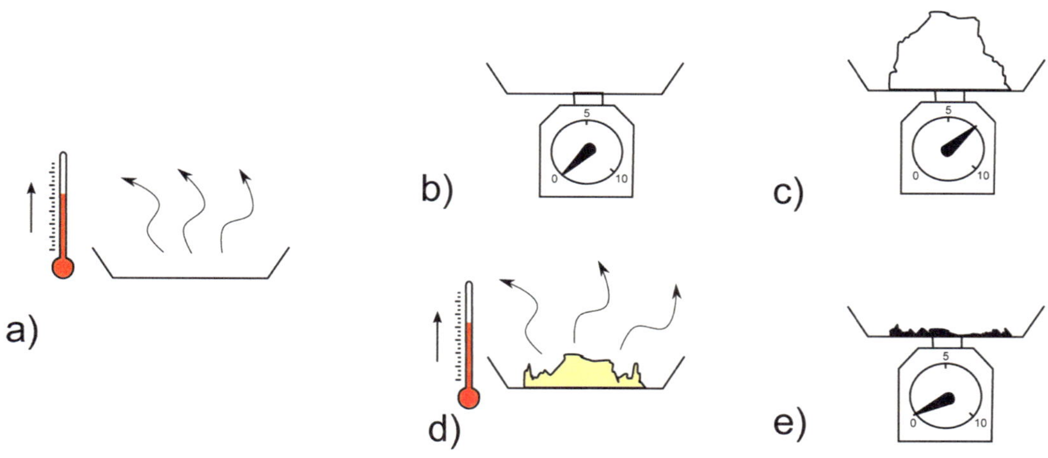

Figure 24. Methodology for determination of residual ash.

In this method, the properties are determined with a test job, just as in the case of the sand casting process. As material system, the PolyPorB system from voxeljet is used. The test job is built on a VX1000 machine from voxeljet. The surface quality is determined based on test blocks. Waxed as well as nonwaxed samples are used.

5.3. Experimental results

The density of the plastic components differs from the density of these parts because of the massively different density in relation to sand components. A mean of 0.63 g/cm^3 is measured .The scattering is relatively small with a range within a build direction of 0.02 g/cm^3. The anisotropy follows the above-mentioned measures, but is less.

Orientation	X			Y			Z		
	Min.	**Mean**	**Max.**	**Min.**	**Mean**	**Max.**	**Min.**	**Mean**	**Max.**
Density (g/cm³)	0.62	0.63	0.64	0.63	0.64	0.65	0.62	0.63	0.64
Modulus of elasticity (GPa)	0.34	0.35	0.37	0.28	0.30	0.37	0.26	0.28	0.29
Ultimate tensile strength (MPa)	3.3	3.4	3.5	2.2	2.6	3.5	1.7	2.0	2.1
Elongation at break (%)	1.4	1.5	1.6	1	1.2	1.5	1.7	0.9	2.1

Table 10. Mechanical parameters of components from the material PolyPorB for use in the investment casting process.

There is a close correlation between the modulus of elasticity and tensile strength for this material system. The tensile strength reaches a maximum of 3.3 MPa and a mean of 2.7 MPa. Here, the relative anisotropy of the strongest to weakest direction is 60%. The elongation at break is consistently less than 1.5%. Therefore, in the fracture behavior, it behaves similar to a glass-like material (refer to **Table 10**).

Orientation	X			Y			Z		
	Min.	**Mean**	**Max.**	**Min.**	**Mean**	**Max.**	**Min.**	**Mean**	**Max.**
Surface quality Ra (μm)	15.9	18.8	22.4	18.1	21.4	25.8	17.2	19.5	22.2

Table 11. Surface characteristics of PolyPorB parts for investment casting.

The length deviation of the material depends in part on the storage and aging of the components. In compliance with the prescribed standard processes, the measurement shows an average percentage deviation from 0.7%. The deviation is always negative, and thus produced shrinkage. The shrinkage is anisotropic and reaches its minimum in the Z-direction. This shows a correlation to the direction of the weakest point.

On average, q roughness Ra of 19.9 μm is achieved. The highest roughness is an Ra of 25.8 μm. If parts are waxed, the roughness varies according to the waxing procedure. If the component is dipped briefly, an Ra significantly below 10 μm is achieved. The components that are dipped longer the roughness corresponds to that of a nonwaxed part (**Table 11**).

The thermal expansion of nonwaxed components largely corresponds to the base material PMMA. The measurement shows a thermal expansion coefficient of $100 \times 10^{-6}\ K^{-1}$. In the region of low temperatures, the wax on the surface of the waxed components will melt. Thus, the thermal expansion seen in the investment casting process can be compensated during the burn-out of the printed component.

The test pieces which were immersed for a short time show a homogeneous image with no color inclusions along the surfaces during the penetration test. The edges of the component exhibit a slightly porous surface, which can be identified due to its pink color. Long immersed parts turn totally pink and penetrant testing confirms the (adverse) reading of the surface condition. In bodies having complex geometries, particularly the outer and inner edges are placed where errors occur. Cracks and tears in the wax surface can be clearly observed by their distinct red color (**Figure 25**).

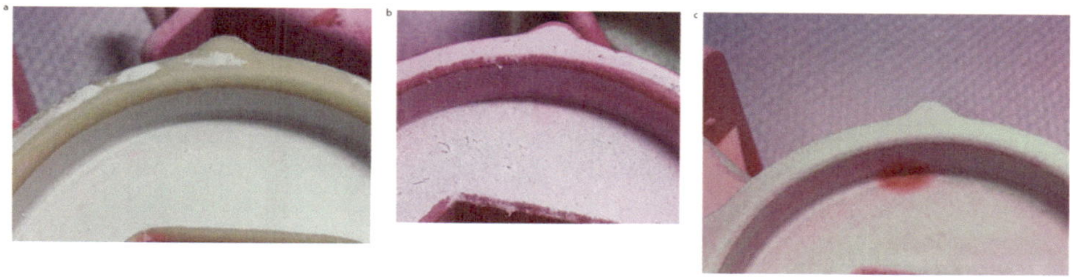

Figure 25. Results of the color penetration test: (a) short immersion, (b) long immersion, and (c) problem areas.

For the material system used, the measurement of the residual ash of samples indicates an ash content of less than 0.3 per thousand based on the initial weight.

5.4. Discussion

In the case of investment casting processes, as opposed to the conventional sand casting methods, the finished models are not directly comparable with PBJ-printed models. In the first process, the part to be compared is built up by injecting a liquid, and in the second process the part is made using particulate material. The strength of PBJ-printed components is sufficient for safe handling during subsequent process steps. The high brittleness needs to be considered during handling. However, the lower density, which is present due to the porosity of the component, makes the parts easy to handle.

The geometry is strongly influenced by shrinkage. However, this shrinkage can be easily simulated and nearly completely compensated by an increase in the geometry of the raw data. The surface roughness of the models in the untreated state does not meet the requirements of a "precise casting procedure". Through a postprocess procedure which is suitably adapted to the task (in this case wax treatment), the desired measurements can be achieved here as well.

The surface of the model has to be sealed, as during the post-treatment process it is immersed in a liquid. The surfaced can be sealed by immersing them in wax. The color infiltration method shows that special care is needed and further steps to ensure that the part is completely coated (sealed) may be necessary.

Thermal expansion of the basic structure of the particulate material can be compensated by applying a coating of wax. The type of wax selected and its melting point play an important role.

During burn-out which exposes the casting cavity, very little residual ash is left behind, this can be attributed to the combination of the raw materials, namely, PMMA and wax used in the process. By applying a suitable process, cleaning of the form can be avoided after burn-out and metal can be poured directly into the hot mold.

6. Summary

Models, e.g., molds and cores, for casting applications can be created using the additive manufacturing process PBJ. Liquid material can either be added directly into the mold after printing to obtain a cast part or a printed model is further processed using the investment casting process. The results show that the material properties meet the demands for a wide application in this field of application. Metals, as well as resins and thermosets, ceramic, or concrete fluids are processed. Quality assurance during production can be assured by applying the testing and measuring methods described in this chapter.

Due to the ongoing development in technology, the production quantities as well as performance can be enhanced. This advancement in technology and knowhow goes hand-in-hand with the reduction in costs, which makes it possible to produce even larger series using the PBJ method. At present, series comprising a few thousands parts are already economically viable. Through this ever-increasing advancement in technology, even larger series will be realized and thus inflexible production machines and dated technologies will eventually be replaced in several areas of production.

Author details

Daniel Günther* and Florian Mögele

*Address all correspondence to: dg@voxeljet.de

voxeljet AG, Friedberg, Germany

References

[1] Ederer, I. (2004). "Werkzeuglose Formherstellung mittels 3D-Drucktechnologie," Giesserei-Praxis 11/2004, pp. 407–410.

[2] Hopkinson, N., Dickens, P. (2001). "Rapid prototyping for direct manufacture," Rapid Prototyping Journal, 7(4), 197–202.

[3] Sachs, E., Haggerty, J., Cima, M., Williams, P. (1989). "Three-dimensional printing techniques," US Patent 5,204,055.

[4] Kashani K., Ederer I. (2004). "Method and device for applying fluids," US Patent, US 8,096,262 B2.

[5] Nassar, A., Reutzel, E. (2013). "A proposed digital thread for additive manufacturing," International Solid Freeform Fabrication Symposium, Austin, TX.

[6] Heinzl. J., Hertz, C. (1985). "Ink-jet printing," Advances in Electronics and Electron Physics, *65*, 91.

[7] Le, H (1998). "Progress and trends in ink-jet printing technology," Journal of Imaging Science and Technology, *42*, 1.

[8] Günther, D., Heymel, B., Günther, J, Ederer, I. (2014). "Continuous 3D-printing for additive manufacturing," Rapid Prototyping Journal, *20.4*, 320–327.

[9] Chua, C., Leong, K., Lim, C. (2010). "Rapid Prototyping: Principles and Applications," 3rd ed . World Scientific Publishing, Singapore.

[10] Hull, C. (1986). "Apparatus for production of three-dimensional objects by stereolithography," US Patent 4,575,330.

[11] Zocca, A., Colombo, P., Gomes, C., Günster, J. (2015). "Additive manufacturing of ceramics: issues, potentialities, and opportunities," Journal of the American Ceramic Society, *98*(7), 1983–2001.

[12] Günther, D., Ederer, I. (2013). "Device and method for 3D printing methods, with accelerated execution," Patent WO 2015/096826 A1.

[13] Fredriksson, H., Akerlind, U. (2006). "Materials Processing During Casting," Wiley, New York, pp. 2–13.

[14] Flemming, E., Tilch, W. (1993). "Formstoffe und Formverfahren," Dt. Verl. Für Grundstoffindustrie, 1, 1st Edition, Leipzig.

[15] Dimitrov, D., Schreve, K. (2006). "Advance in three dimensional printing – state of the art and future perspectives," Rapid Prototyping Journal, *12.3*, 136–147.

[16] Budzik, G. (2007). "Possibilities of utilizing 3DP technology for foundry mould making," Archives of Foundry Engineering, Foundry Commission of the Polish Academy of Sciences, *7*(2), 65–68.

[17] Wilkening, C. (1997). "Fast production of technical prototypes using direct laser sintering of metals and foundry sand." In: Proceedings of the 2nd National Conference on Developments in Rapid Prototyping and Tooling, Buckinghamshire, UK, pp. 18–19.

[18] Campbell, J. (2011). "Complete Casting Handbook: Metal Casting Processes, Metallurgy, Techniques and Design," Amsterdam, 1st ed. Elsevier.

[19] Dawson, M. (2001). "Silica sand- foundry requirements and classification," Metal Casting Technologies, *47*(4), 24.

[20] Recknagel, U., Dahlmann, M. (2008). "Spezialsande-Formgrundstoffe für die moderne Kern-und Formherstellung," Giesserei-Praxis, (11), Vol. 11, pp. 401.

[21] Spath, S., Seitz, H. (2014). "Influence of grain size and grain-size distribution on workability of granules with 3D printing," International Journal of Advanced Manufacturing Technology, *70.1-4*, 135–144.

[22] Blais, F. (2004). "Review of 20 years of range sensor development," Journal of Electronic Imaging, *13*(1), p231–240.

[23] Nawy, E. (1997). "Concrete Construction Engineering," 1st ed. CRC Press LCC, Boca Raton.

[24] Peters, J. (1991). "Practical Timber Formwork," 1st ed. E & FN Spon, London.

[25] Cheah, C.M., Chua, C.K., Lee, C.W., Feng, C., Totong, K. (2005). "Rapid prototyping and tooling techniques: a review of applications for rapid investment casting," The International Journal of Advanced Manufacturing Technology, *25*(3–4), 308–320.

[26] Cowie, J. (1991). "Polymers: Chemistry and Physics of Modern Material," 2nd ed. Chapman & Hall, London.

Section 2

Medical Applications of 3D Printing

Chapter 4

Colour Image Reproduction for 3D Printing Facial Prostheses

Kaida Xiao, Sophie Wuerger, Faraedon Mostafa,
Ali Sohaib and Julian M Yates

Additional information is available at the end of the chapter

http://dx.doi.org/10.5772/63339

Abstract

In this chapter, using colour 3D printing technology, a 3D colour image reproduction system is detailed for the semi-automated and accurate additive manufacturing of facial soft tissue prostheses. A protocol for 3D colour image reproduction was designed based on the six steps of processing. For this specific application, protocols for each sub-process required development and details of each technique applied are discussed. The quality of facial prostheses was evaluated through objective measurement and subjective assessment. The results demonstrated that the proposed colour reproduction system can be effectively used to produce accurate skin colour with fine textures over a 3D shape, with significant savings in both time and cost when compared to traditional techniques.

Keywords: facial prostheses, colour image reproduction, 3D colour printing, 3D image acquisition, 3D image processing

1. Introduction

Maxillofacial prosthetics or anaplastology refers to the specialty that designs and manufactures prostheses used to replace part or all of any stomatognathic and/or craniofacial structure. The process provides descriptive evidence of the prosthesis, including location, retention, support, time, materials, and form. It is both an art and a science of cosmetics, anatomy and functional reconstruction, that is achieved by means of artificial substitutes of head and neck structures that are missing or defective. It is the branch of dentistry that rehabilitates intra- and extra-oral deformities [1]. Extensive tissue loss involving facial (or indeed any other body structure), on many occasions, cannot be corrected surgically due to a lack of sufficient or adequate donor tissue that may be required for the surgical repair and reconstruction. Furthermore, the patient's age and general condition may not permit extensive surgical procedures or allow for the morbidity often associated with protracted courses of reconstructive surgery and recovery [2]. Additionally, the final outcome can often be aesthetically and functionally compromised. In such cases, defects may be replaced artificially with the provision of facial prostheses to provide functional rehabilitation and aesthetic improvements. Often there is an associated improvement in social, emotional status and overall quality of life [3].

Traditional methods of prosthesis production are well established and are used even today. These include taking an impression, manufacturing a cast and ultimately hand crafting a prosthesis in a silicone based or similar material. The provision of prostheses in this manner has provided considerable comfort and support to many patients over many years, and allowed them to continue with normal day-to-day activities and enhance their social interaction [4]. Despite the advantages this method can provide, its application has shown some limitations and shortcomings. These are primarily related to the processing strategy, technical expertise required, time, effort, cost and retention problems. Furthermore, there are durability issues due to material degradation and colour fade after a relatively short period of service because of general wear and tear and exposure to ultraviolet radiation. For these reasons, facial prostheses require renewing and periodic replacement, which is a costly and time-intensive process, which places a burden on both patients and prosthodontists alike [3].

In the last decade, additive manufacturing technology, including three-dimensional (3D) printing, has advanced dramatically. Colour 3D printing has also evolved to produce full spectrum coloured solid objects utilising a range of materials [5, 6]. With the evolution of various 3D imaging techniques, accurate acquisition and transformation of target object geometric data into 3D digital models can be achieved. By combining the 3D image capture and printing techniques, there is huge potential to achieve 'What You See Is What You Get' processing. More importantly, it has the ability to directly interconnect with advanced manufacturing techniques, allowing customisation with high accuracy, resulting in savings of both time and costs [7]. It has been extensively utilised in rapid prototyping [7], successfully applied in medical sciences [8, 9], and is gaining popularity in multidisciplinary applications [10–12]. For medicine, captured digital 3D models have great accuracy and have been effectively used for facial disfigurement diagnosis, surgical planning and assessing treatment outcomes for several years [13, 14]. Additionally, there is the potential to develop this automatic additive manufacture technology for facial prosthetics.

Compared with conventional image capturing technology, image processing between 3D image devices has much more complicated working processes. For 3D printing, the quality of 3D printed objects is not only affected by the printing itself including binder/substrate interaction and printer resolution, but also the printing material and any post-processing or finishing stages [15]. Therefore, without a specific protocol, 3D objects can often be produced with poor reliability, accuracy and quality. Moreover, in terms of 3D image reproduction, image processing methods to transform 3D images from a 3D camera to a 3D printer are far less well developed than existing processes that use 2D technology. Furthermore, accurate colour reproduction for facial prostheses is highly desired and the quality of skin colour reproduction can affect the overall quality of the facial prostheses significantly. Therefore, accurate colour management processes are not only essential, but also needs to run in conjunction with the specific 3D manufacturing processes used.

In recent times, new and innovative methods for manufacturing facial soft tissue prostheses that prioritises accurate 3D colour image reproduction have been developed and a framework and protocol for specific 3D processing designed [16–18]. Additionally, colour management processes have been developed and successfully applied to 3D imaging devices and manufacturing processes.

2. Protocol of colour image reproduction

Based on the steps shown in **Figure 1**, a 3D colour image reproduction protocol has been developed for the automated 3D printing of facial prostheses. The first step involves 3D scanning of the face using a 3 pod 3dMD photogrammetry system (3dMD, Atlanta, GA, USA). The system captures both 3D topography and colour information of the scanned surface in controlled illumination and viewing conditions. The resulting mesh data from the scanner often has errors in geometry and colour and therefore requires further editing before it can be sent to the printer. Using a combination of software suites, for example Magics (Materialise, Leuven, Belgium), the raw scanned data can be modified and corrected by removing noisy polygons, along with colour adjustment of areas with specular highlights or other inaccuracies. In order to add realism, the next step involves the addition of fine texture (e.g. pores, wrinkles) over the 3D mesh using high-field mapping. Finally, thickness is added to the mesh data in order to get a solid printable model.

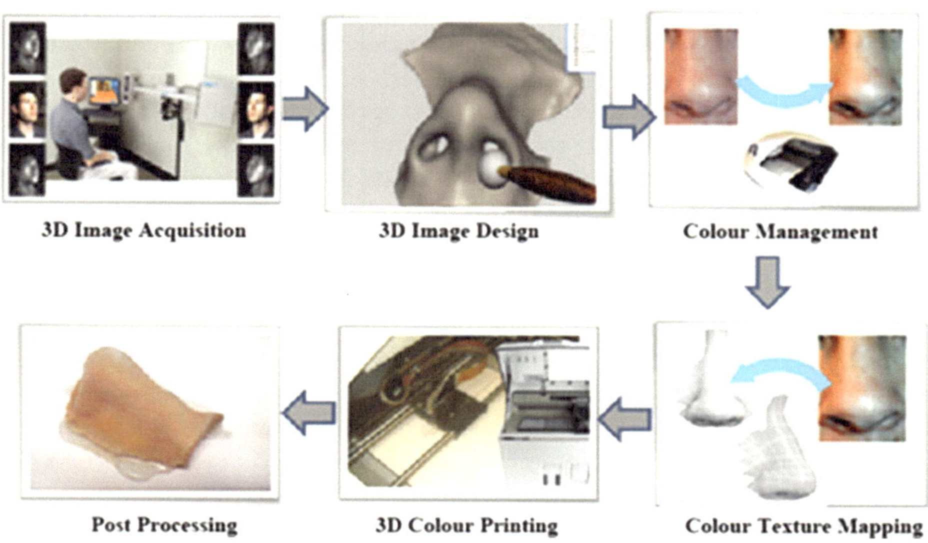

Figure 1. 3D image reproduction protocol.

As with the surface topographic information, the colour images from the 3dMD camera system may also require further processing before they can be overlaid on the 3D mesh prior to final colour printing. This primarily involves colour management of the 2D colour image from the camera RGB to printer RGB for each pixel using specific camera and printer colour profiles respectively. For this specific task, the colour profiles used were developed using both conventional colorimetric and spectral-based reproduction methods [19, 20]. When the colour management is finalised, surface texture mapping is conducted to map the newly generated colour image onto the manipulated 3D model. The penultimate step involves 3D colour printing to produce the 3D model using the 3D printing system, and then, finally as a post-processing step, the strength and flexibility of the printed model are improved by infiltrating it with medical grade silicone.

3. 3D image acquisition

3D facial data can be obtained using a variety of different techniques. CT and MRI scans have been used for the RP of internal body organs and structures as well as facial prosthesis prototyping. These techniques have an advantage over external body 3D scanning as they can be used for generating patient specific 3D models of internal body parts or organs, etc. However, due to their lower resolution, they are unable to capture fine textural details like pores and wrinkles on the surface of the body. Furthermore, data capture is undertaken in supine positions and thus can affect the accuracy of captured data in that the positional detail is not truly reflective of that in a natural upright repost position. Finally, there is no ability to capture colour data using these modalities.

Compared to CT and MRI, 3D photogrammetry is becoming more popular due to the expediency of the data capture, lack of exposure to ionising radiation, ability to capture data in a prone position, and relatively easy storage, transfer and utilisation of the data. Furthermore, the cost of such systems is relatively affordable. Active 3D scanning techniques like laser scanning [21], Kinect [22], structured light [23] and speckle projection stereo [24] have also been used previously to obtain 3D scanned data for manufacturing of facial prosthetics. Although the image acquisition and 3D reconstruction techniques are different using these methods, the outputs produced are similar in terms of the data produced and they all produce 3D mesh data and colour information for the scanned geometry.

Typically surface scanners, including the 3dMD scanner, are based on the speckle projection technique and was used successfully in this project. This system is based on stereo photogrammetry and calculates stereo correspondence by projecting infrared speckle patterns over the target surface to then create a 3D depth map of the scanned surface. It consists of several machine vision cameras and infrared speckle projectors combined with a flash system to acquire texture and 3D topographic information simultaneously. These are collectively known as pods. The process then merges all the viewpoint data from each pod to create the final 3D polygon mesh with the colour images then laid over the 3D mesh to add colour, texture and appearance. The resulting mesh can then be saved as an .OBJ file for processing and editing purposes. An example of this scanned data is shown in **Figure 2**.

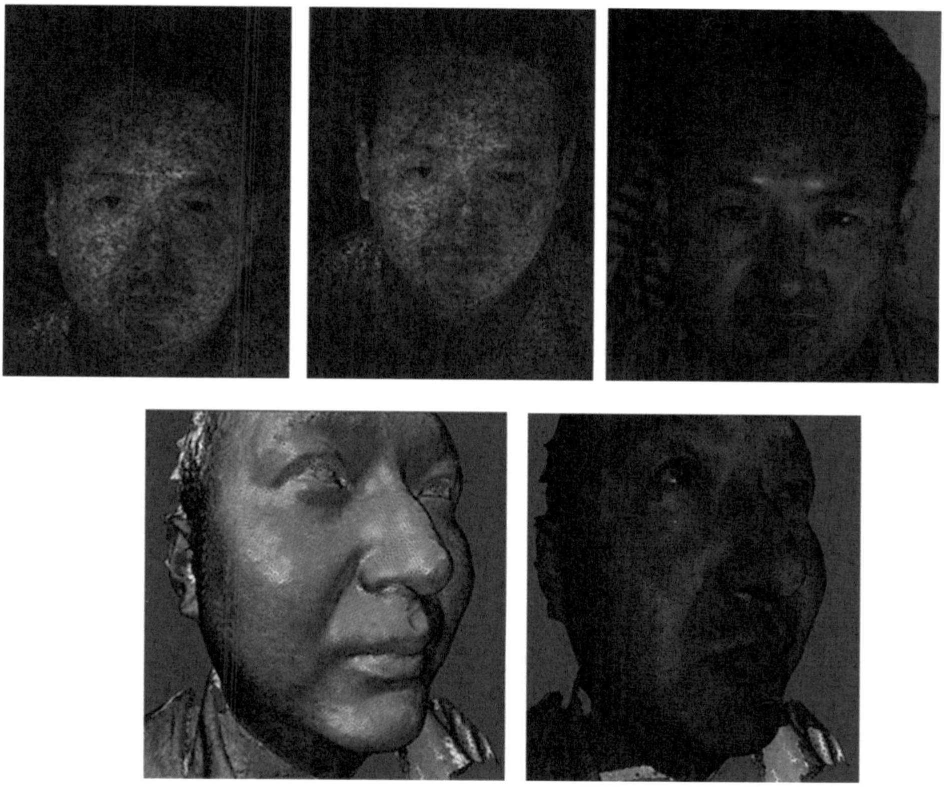

Figure 2. (a)–(c). a) Speckle projection images and texture bitmap captured from a single pod. (b) 3D mesh created 3 using all viewpoint data. (c) The 2D bitmap image overlaid over 3D mesh to add colour.

To ensure sufficient quality images, protocols can be developed to aid data capture. These can include—capturing images in a blackout facility, minimising patient movement, employ consistent peripheral lighting protocols and following the camera manufacturers recommended protocol for image capture, including geometric and colour calibrations immediately prior to imaging. When capturing images to use in the 3D printing of facial prostheses, patients should be asked to remain motionless during the image capturing process in order to avoid any motion artefacts. If the patient has already had surgery to remove the affected facial part, they will often have precision attachments retaining existing prostheses. In cases such as these, due to the limitations of data capture for these attachment—small componentry, shiny surfaces, obscured by anatomical features including prominences, undercuts and recesses, acrylic bosses (described below) can be used to ensure the camera captures the exact location of any precision attachments being used. Ideally, the patient should also be maintained a neutral pose during imaging.

4. 3D image design

Raw data obtained from 3D camera systems often require further processing prior to 3D printing. Depending on the anatomical location of the prosthesis, mesh editing can consist of the following steps.

4.1. Data preparation

CAD generation of the missing facial part can take various forms. Data capture of unaffected areas of the face can be undertaken and images 'mirrored' in order to virtually replace those affected or missing. This may be suitable for missing ears or orbital areas where there are usually bilateral structures. However, this may not be appropriate or possible for single or isolated structures including nasal structures or where surgery has already been undertaken to remove the affected part. Alternatively, open-source CAD databases are now increasingly available—www.turbosquid.com—and these can offer CAD users direct links to downloadable content. These may include non-specific facial shapes covering a wide range of anatomical form, from those considered 'generic' to those with interesting or characteristic features. Although easily accessible, there are some issues with these sources. Although geometrically accurate, the peripheral area can be quite variable and not suited to cover the affected areas in specific patients. Additionally, the surface detail can be highly faceted and unsuitable for direct modelling. However, these files can provided good starting points for CAD design and allow adaptation to enable conformation to the affected area. Various software can be used to alter theses captured or acquired data files including 3-Matic (Materialise, Leuven, Belgium). These can then be used to alter the surface detail and to refine the geometry in order to provide individual and bespoke characteristics for the missing facial part (**Figure 3**).

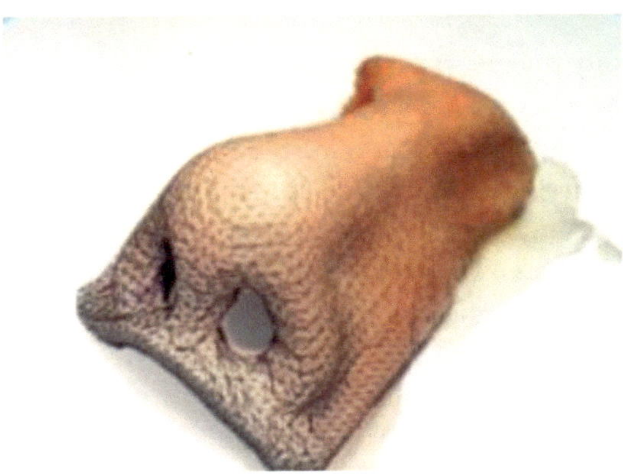

Figure 3. Image detailing manipulation of the CAD prosthesis to refine the shape and fit of the final part (Approx. part Dim: 40 × 24 × 25 mm).

4.2. Data manipulation

The acquired and adjusted model can then be blended into the affected area to create a smooth margin that will contact the facial tissues. Furthermore, CAD prostheses occasionally require hollowing and thinning until the prosthesis is between 2 and 3 mm thick and this can again be undertaken with various software packages. Finally, the model is then trimmed to remove all unnecessary data prior to printing. As previously indicated, information for attachments can then be imported and merged with the prepared CAD prosthesis and be precisely aligned to produce a good retentive element and fit.

4.3. Precision attachment design

Often facial prostheses are retained using precision attachments including osseointegrated implant retained magnets, bars and clips. In these situations, external markers, or bosses, are

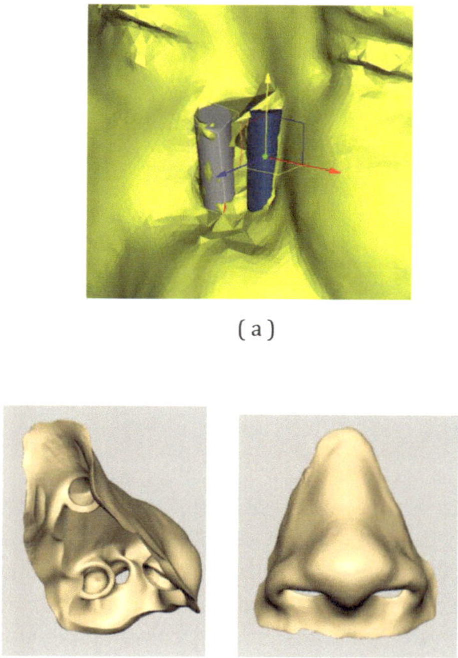

(a)

(b) (c)

Figure 4. (a) Images illustrating the use of external markers, or bosses, to detail the position of precision attachments to aid retention of the prosthesis. (b) Translation of the geometric position of the attachment points for the precision attachments within the prosthesis (Approx. part Dim: 40 × 24 × 25 mm). (c) Checking of CAD design to check no interference of surface geometry caused by incorporation of precision attachments.

used to identify the position, angulation and orientation of any attachments that need to be incorporated into the data incorporated in to the process. Their position can then be translated through the design and manufacturing process into the final prosthesis (**Figure 4**).

4.4. Data detailing

Natural orifices such as nasal apertures and auditory canals must be designed to ensure that the prosthetic has correct anatomical form and be modelled in a way that will allow for appropriate function if needed. For the nose, it will allow for the passage of air during respiration whilst for the ear, although external form may be missing, it will allow non-impeded hearing via direct access to the external auditory meatus (canal). Similar considerations can be made when designing orbital prostheses as the housing of an artificial glass or acrylic globe will be required within the palpebral fissure (eye lids). In some instances, it should be remembered that the provision of attachments may impinge on these design features and so optimal modelling and detailing must be undertaken to camouflage these issues.

4.5. Design finalisation

Once the final model has been produced, additional details need to be added in order to produce a satisfactory prosthesis. Final details including the incorporation of peripheral extensions, feathered edges, any necessary strengthening buttresses and the smoothing of seams and joins on the model's inner surface can all be undertaken prior to printing. The colour calibration image is then also applied to the whole model prior to production.

5. Colour management

Due to the different ways in which colour communication between 3D image capture and 3D image printing is utilised, there are often significant discrepancies in colour between the printed and original objects when using this type of manufacturing process. Conventional colour image reproduction techniques based on CIE colorimetry have been used for more than 80 years and perform very well in transforming colour images from one digital media to another under various viewing conditions [25]. However, the application of conventional colour image reproduction techniques for 3D printing technology is not as straightforward. With this protocol, 3D colour objects can be split into a monochromatic 3D objects and a 2D colour images. With this in mind, a conventional colour reproduction technique can be applied to transform the 2D colour image from camera RGB to the corresponding printer RGB through human colour appearance attributes. To achieve an accurate colour reproduction through 3D imaging devices, specific colour profiles need to be developed in order to link the colour system of a specific device to the human visual system [26].

5.1. Development of colour profiles for 3D image devices

To truly reproduce colour from a 3D camera to 3D printer, both a camera colour profile and a printer colour profile need to be developed to connect camera RGB and printer RGB with human eye response (CIE XYZ tristimulus values [27]). Variations in skin colour and tones are considerable [28], and to try and incorporate the whole spectrum of skin colours available for different human populations templates such as the digital Macbeth ColorCheckerDC charts (X-Rite Inc., Grand Rapids, MI, USA) can be used to provide numerous training colours (see **Figure 5a**). Utilising this method, a 2D colour chart can be converted to a 3D model with dimensions of 200 (l) × 150 (w) × 3 (h) with the colours desired and printed using a 3D printer such as a Z Corp Z510 colour printer. These can then be used as your testing colours (**Figure 5b**). Subsequently, the CIE XYZ tristimulus values for each training colour produced in the printed colour chart can be obtained by taking colour measurements using a spectrophotometer such as a Minolta CM-2600d (Konica Minolta Inc., Tokyo, Japan). During measurements, standard viewing conditions should be applied such as a viewing geometry of d/8 (diffuse illumination, 8-degree viewing), with specular component included and aperture sizes consistently set to a defined diameter—3 mm. The illuminant must also be consistent and ideally of an industry standard including setting to CIE standard D65 to simulate skin colour in daylight conditions and CIE 1931 standard observer [27].

Figure 5. 3D models of a) training and b) testing chart.

Based on printer RGB and CIE XYZ tristimulus values for the training colours, a printer colour profile can be developed using a third-order polynomial regression model [28]. A camera colour profile can also be produced utilising the same chart. Colour images are captured by the 3dMD camera system and the camera RGB for the number of training colours used identified. Then, based on these camera RGB and CIE XYZ tristimulus values, a camera colour profile can be developed using a second-order polynomial regression [29]. Using this method, for each pixel of image, camera RGB is first transformed to CIE XYZ tristimulus values and then transform back into printer RGB.

5.2. Evaluation of colour reproduction

To evaluate colour reproduction for the human face, a colour test chart first needs to be designed. This can be undertaken using 14 predetermined human skin colours, including four Caucasian, two Chinese, two Asian, four African and two Caribbean skin shades (see **Figure 5b**) [30]. A 3D colour chart must then be generated using the colour printer to be used (Z Corp Z510), with defined and consistent dimensions. After post-processing in the same way any final prosthesis would undergo, this chart is referred to as the original colour chart.

Colour reproduction can then be evaluated using two reproduction charts that must be produced using two different 3D colour image reproduction systems. When the first chart is produced, the colour image for the original chart must be captured using the 3dMD camera system and then processed with only minor corrections in 3D geometry before it is sent to the Z Corp Z510 printer for 3D printing. This printed colour chart can then be referred to as the first reproduction (Chart 1). For the second processing, it can be conducted following the proposed 3D colour image reproduction process. This printed chart can then be referred to as the second reproduction chart (Chart 2).

To assess the performance of the colour reproduction, CIE XYZ tristimulus values for each colour patch within each colour chart must be measured using a spectrophotometer (Konica Minolta cm-2600d). Colour difference between the original chart and each of the two reproduction charts for each of the 14 testing colours under CIE illuminant D65 should be calculated using a CIELAB colour difference formula [31]. The mean, maximum, minimum and standard deviation for the colour difference can be recorded and tabulated. Examples of these values can be seen in **Table 1**. If undertaken correctly, this method will demonstrate that a significant improvement in colour reproduction can be achieved using 3D colour image reproduction systems. For its successful application in the production of facial soft tissue prostheses, an acceptable colour difference for the 3D printed objects is approximately 3–4 ΔE^*_{ab} [32].

CIE ΔE*ab	Mean	Max	Min	SD
Org vs. Rep1	20.8	27.8	8.0	5.5
Org vs. Rep2	4.5	11.1	2.5	2.3

Table 1. Examples of colour difference between original (Org) colour chart and reproduction (Rep) colour charts.

6. Colour texture mapping

Many 3D photogrammetry systems can give the illusion of texture by wrapping the 2D image over the 3D surface. This wrapped texture does not actually produce fine wrinkles, pores or blemishes on the surface of the skin and hence cannot be reproduced within the rapid prototyping process.

Figure 6a demonstrates a side-by-side view of a 3D polygon mesh and the 2D bitmap image captured by a 3D camera system (3dMD System). It can be seen from the image that the 3D mesh does not actually contain fine details including pores and wrinkles, which are clearly visible in the 2D bitmap. Each polygon in the 3D mesh is linked to a particular region in the 2D bitmap. **Figure 6b** shows highlighted polygons in the 3D mesh and the region used from the 2D bitmap to overlay colour onto a monochromatic polygon.

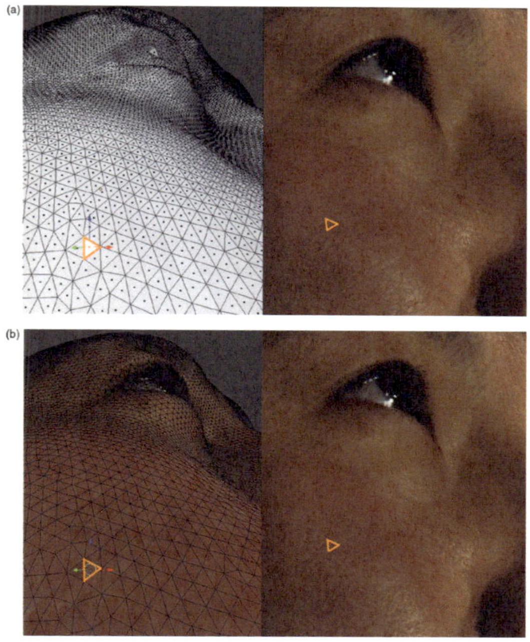

Figure 6. (a, b). Images demonstrating a 3D polygon mesh and the 2D bitmap image captured by a 3D camera system.

To enhance realism and improve the characterisation of the patient's skin, various techniques have been developed. One such method includes incorporating surface details such as pores, wrinkles and fine lines into the 3D model. Height-field mapping (also known as bump mapping) within the CAD design process can be used to translate the texture reference image onto a geometric pattern. Height-field mapping is a method of translating fine 2D images onto 3D geometric virtual models. This is based on white (high) and black (low) greyscale texture reference mapping as detailed above, and can be applied to map representative texturing to the appropriate areas of the facial model. These techniques can not only use patient specific data from adjacent facial anatomy/topography but also, using pre-treatment 2D photographs of the same area. The latter allows for manually adding realistic skin textures to the prosthetic model surface in cases where texture from other parts of the missing face are either inappropriate or not consistent or representative of the area to be replaced. Based on the grey level intensity (from white to black), computer software can control the depth of imperfections over a skin surface. An original skin depth map is shown in **Figure 7 (a)** and can be used to add texture over a flat surface. Surface topography changes according to the grey level intensity and the positioning of individual pores and wrinkles is shown in (b). The resultant mesh can be 3D printed, producing a skin-like texture over a flat surface (c).

Figure 7. (a) Depth map of skin showing pores and fine wrinkles as dark areas. (b) Software-induced texture mapping over a flat surface using the depth map from (a). (c) 3D printed surface with pores and wrinkles clearly visible on the surface (Approx. part Dim: 38 × 38 × 3 mm).

Given the flexibility of such software, texture can be mapped not only onto flat surfaces but also over complex 3D shapes. An example of this texture mapping over a nose model with varying pore depths is show in **Figure 8**.

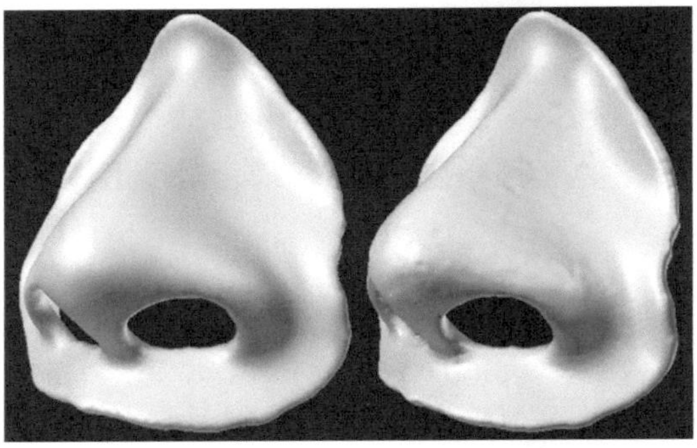

Figure 8. Illustrations detailing texture mapping. (Approx. part Dim: 70 × 61 × 29 mm).

However, the detail provided in the final prosthesis is dependent not only on the resolution of 3D data obtained but also the resolution of the 3D printer and characteristics of the powder and binder type used within the process. The use of course powder in the printing process will reduce the detail of the texture derived from the height-field mapping even though such detail can be mapped within the CAD process. In contrast, using finer powder will enable the addition of very fine details over the printed prosthesis.

7. 3D colour printing

Full colour 3D printing is considered advanced technology with different 3D colour printing technologies continually being developed and evolving. One common full colour 3D printing technology is 3DP™ printing, also known as powder–binder printing. Developed at Massachusetts Institute of Technology [33] and licensed to Z Corporation and 3D Systems, the process itself is based on inkjet printing, with the powder being deposited in consecutive layers, which are then selectively joined by ink-jetting with coloured binder. Three (CMY) or four (CMYK) coloured binders together with a clear binder are mixed to print powder material in a full colour spectrum, layer by layer. After 3D printing, post-processing, including the removal of excessive unused powder and infiltration, often needs to be conducted in order to produce the final model (**Figure 10**). The powders can be made of different kinds of materials. Gypsum is primarily used in combination with plastic powder. However, starch, ceramic, glass and other powdered materials can also be processed as well. For manufacturing facial prostheses, 3D printers including the Z Corp Z510 3D printer (3D Systems Inc., Rock Hill, SC, USA) can be used to print colour into biocompatible starch powder. Post-printing processing is then required. Infiltrating with suitable elastomeric polymer can then be undertaken to produce a flexible, lightweight and lifelike soft tissue prosthesis. More recently there has been the development of other elastomeric 3D printing processes including direct deposition and filament printing. However, their main drawback is the limited spectrum of colour they can print in.

7.1. Infiltration base design

The printing process itself will produce a highly accurate solid model of the missing facial part and consists of printed powder and skin coloured binder. Given the resultant structure post-printing processing is required to infiltrate the model with a suitable material with properties that will lend itself to the role required—light, flexible and durable in nature [34]. Due to the post-printing processing needed for direct prosthesis printing, an infiltration base modelled from the affected and adjacent area must be generated (**Figure 9**). The site of the affected area is imported into the process as a single surface using a suitable CAD software programme. Then using the final prosthesis as a guide, the infiltration base can be trimmed to support the prosthesis and any peripheral extension required in order to define the contact point or provide a feathered edge for final fitting. Each requirement of the prosthesis can be developed within the support with the exact nature of the periphery dependent on the type of facial part to be replaced and the characteristics of the surrounding tissue. Smooth peripheral surfaces will be more suited to feathered edges whilst well-demarcated wrinkles, or surgical margins will be more suited to a well-defined edge for closer fitting. Final design features in the infiltration base will depend on the geometry of the missing facial part, but drainage holes and reservoirs may be required in order to provide adequate support and infiltration.

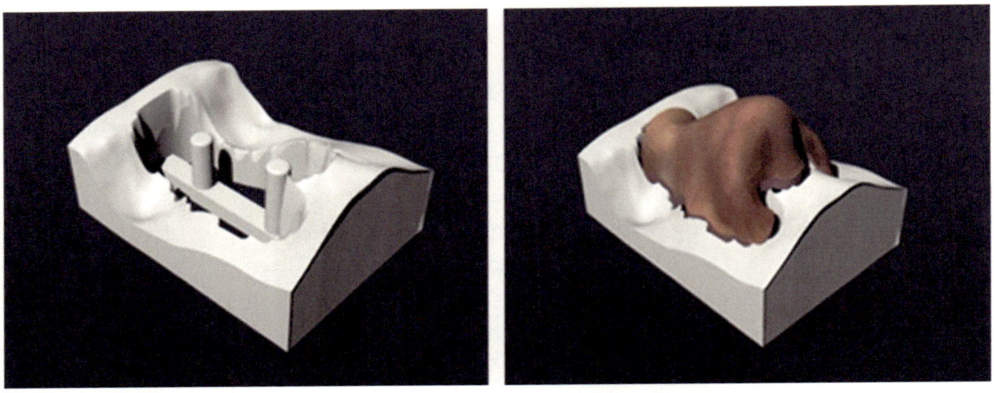

Figure 9. Illustrations demonstrating the design and production of an infiltration bed to support the prosthesis during infiltration with a suitable material (Approx. part Dim: 52 × 36 × 25 mm).

7.2. Manufacturing

Following CAD manipulation of the model and design of the final prosthesis, data can be converted to .zpr files for exporting to the printer. Again this can be undertaken using suitable software. Using layered production with biocompatible powder and coloured binder, the prosthesis can be printed ready for any post-processing including infiltration with a suitable elastomeric polymer. In many cases, the resultant printed model is required to rest for a period of time to allow complete drying of the binder and gentle removal of excess powder allows for accurate post-processing. As detailed previously, it may well be that for large models that have been designed to replace significant portions of facial tissue an infiltration base is required to prevent distortion or sagging during this process. This can be manufactured separately using any suitable hard materials including acrylic, plastic or even metal either by printing or milling.

7.3. Prosthesis infiltration

3D printing produces models with excess powder intimately associated with the final structure. The resultant powder has to be gently removed prior to infiltration with a suitable material. The prosthesis can then be immersed into the infiltrant either with or without the infiltration base and a period of time allowed for complete saturation. This process will be dependent on the thickness of the printed model and viscosity and setting characteristics of the infiltrant. All will have to be evaluated prior to production of the definitive prostheses. Ideally, materials with characteristics and properties close to those used in existing production methods should be utilised in order to make a comparable prosthesis. Normally, the materials employed would be either silicone, acrylic or polyvinyl in nature. Given the variability in the characteristics of these materials extended periods of immersion or infiltration under pressure may have to be considered. Following this process, excess infiltrant needs to be removed from key areas such as the apertures, orifices and attachment housings, and then allowed to fully cure and dry. This process normally takes approximately 24 h.

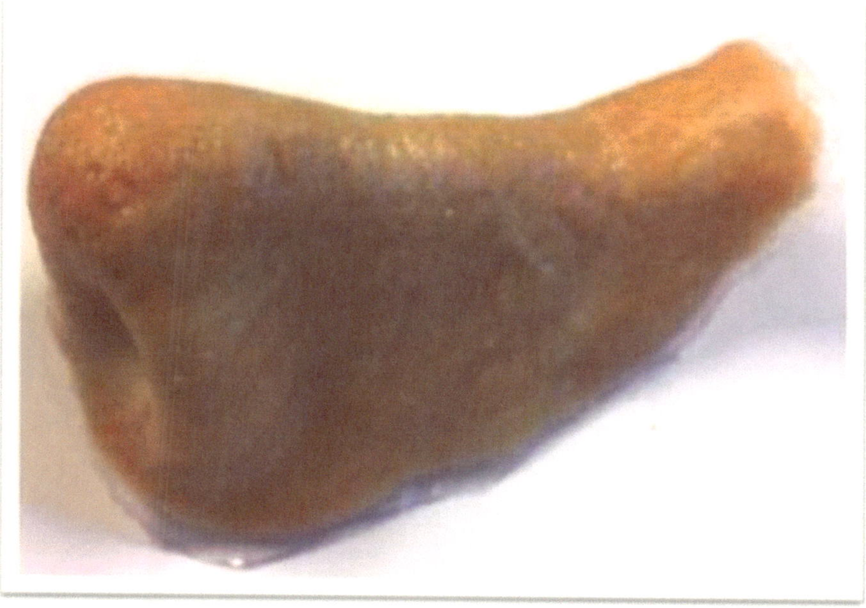

Figure 10. Photograph of a 3D printed nasal prosthesis infiltrated with a compatible medical grade silicone infiltrate (Approx. part Dim: 40 × 24 × 25 mm).

8. Final preparations for patient

Following production of the prosthesis, it can be delivered to the anaplastologist for final fitting. Given the nature of the manufactured periphery and infiltration with a comparable material, adjustments can be made to both the peripheral fitting surface and colour in order to achieve an exact fit and colour match in normal repose and function. In addition to these changes, attachment components may be secured into place within the prosthesis using appropriate adhesives. Finally, minor changes can also be made to the texture and surface appearance with the addition of mattifying agents or external colourings. The latter will allow for a more natural finish. The prosthesis can then be fitted to the patient. **Figure 11** demonstrates a 3D printed nasal prosthesis retained by magnet attachments. Using this contactless approach, production and final delivery were achieved within 48 h. Given that any single print can produce up to 60–80 prostheses in a single run, the relative time required to produce facial prostheses for multiple patients can be reduced significantly.

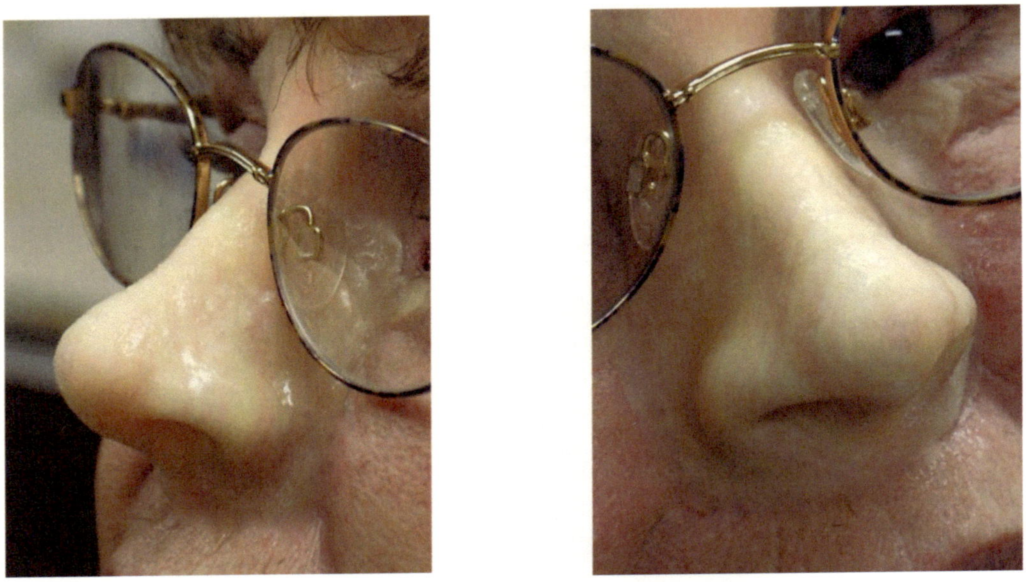

Figure 11. Clinical photograph shows the fitting of a 3D printed nasal prosthesis retained by magnet attachments.

9. Discussion and conclusion

In this chapter, it can be demonstrated that using modern technology and applying some of the traditional methods of the maxillofacial anaplastologist, it is technically feasible to rapidly and directly manufacture realistic soft tissue prostheses using 3D printing. Using available 3D imaging techniques, which are more accurate and convenient than taking impressions, and less harmful to the patient than CT or laser imaging, it can be shown that it is possible to generate a precise life like 3D model of a traumatised or defective area. Furthermore, by incorporating attachments, a prosthesis of exceptional fit and retention can be produced.

To achieve a lifelike and naturally looking prosthesis, any method of 3D printing would also have to incorporate accurate colour management processes, which converts patient's skin tone into a machine-readable format for 3D printing, and provide an accurate intrinsically coloured prosthesis that matches the patient's own skin colour. Furthermore, if suitable infiltration materials are utilised then extrinsic colouring can be added to the prosthesis, either through a coloured infiltrant or painted on by the prosthetist. In addition to the colour management, it is possible to generate a two-part texture design workflow that allows production to accurately match the patient's original features. Using automatic height-field mapping and manual carving to enhance the features, it is possible to generate a prosthesis that resembles the patient's original morphology and skin texture. Given the materials that can be used to print such models consideration must be given to the accuracy that can be achieved, not only in terms of surface texture, but also optimisation of the infiltration protocols. These will inevitably improve the production process and the overall finished look.

To enhance the peripheral fit of the prosthesis, consideration must also be given to the site of fitting and protocols developed that will allow the production of a prosthesis with a durable seal and thin feathered edges or well-defined edges where appropriate. The latter points may be determined by the condition of the existing site, surgery the patient has undergone or the preference of the prosthetist. Given the accuracy of the process and the flexibility to produce either type of periphery, the protocols developed would be suitable for both adhesive or attachment retained prostheses.

There are several significant advantages to the process described in this chapter. By printing the prosthesis directly using biocompatible materials, several steps can be eliminated from the traditional process, including impression taking, mould making, setting the prosthesis in silicon and intermediate fitting steps. Current methods require two to four patient visits over several weeks and significant man hours to produce the final prosthesis. Replacements must use existing moulds or the process must be restarted, and given that old moulds may not be accurate due to changes in disease state or other patient factors, it may not be appropriate to use these for replicating the prostheses. Furthermore, the method described is contactless apart from the final fitting of the prosthesis. As a result, the patient experiences minimal discomfort and inconvenience and data capture is quicker than traditional methods. Furthermore, it enables the prosthetist to store this data electronically for future use and record keeping. Using this digital process, the imaging and colour reference appointment takes approximately 10–15 min. Additionally, the production time is significantly reduced to approximately 48–72 h between imaging and fitting of the finished prosthesis, and as detailed, this process has the potential to produce multiple parts (60–80) in the same timescale and thus reduce the relative production times for each prosthesis significantly. The process is only limited by the CAD input (i.e. image processing and model design/manipulation) although the process is also largely automated. CAD programs can either be used separately, or integrated into one bespoke software that seamlessly passes the model between programs. The only human intervention occurs at the CAD input, design and post-processing infiltrant stages.

One other benefit of developing and using this technology is the significant cost benefits. The average cost to the health care system to produce a soft tissue prostheses in the UK using traditional methods is approximately £2000–£6000, and the cost remains largely the same for each replacements prosthesis. The per-unit cost of 3D printed parts with attachments is significantly reduced due to the reduction in labour costs as well as the numbers that can be produced at any one time. Given that the average prosthesis has a lifetime of 1–2 years, and there is an ever more aging population, the long-term time and cost savings would be significant using this new technology.

However, there are some limitations to this methodology. The process would be dependent on practitioners acquiring new skills, namely software and CAD usage, and this again will be highly specialised. However, in recent years, this technology has been available for conventional hard tissue and dental prostheses, for example, custom-made dental crowns, bridges, implant abutments and various other hard tissue prostheses. Given that the software associated with these developments has been widely accepted and is considered user friendly and intuitive, the introduction of such technology for soft tissue prostheses should not be difficult. Another consideration may be the limited availability of virtual CAD models that may be required if existing patient data did not exist. However, these are becoming more available and given that once patients have had parts made existing electronic data can also be used for repeat prostheses for the same patient or adapted for new patients. Another limitation may be the initial start-up costs, which are not insignificant. At present, equipment costs are high, however, with the advancement of new printing techniques and the emergence of numerous manufacturers in the market place, costs are dropping significantly. Furthermore, collaborative or 'hub and spoke' arrangements may mean that the data capture and manipulation of missing parts could be undertaken at a local/regional level, whilst the manufacturing process could be centralised to a few specialist production centres. Given that the data will be stored electronically, electronic communications are largely effortless, and similar protocols are seen with other manufacturing industries, this should be easily achievable.

In conclusion, the utilisation of modern manufacturing technologies including 3D printing can provide a quality product quickly and at a significantly reduced cost, labour and patient inconvenience. As detailed, this is a viable method for manufacturing prosthesis using commercially available equipment and software and could easily be implemented clinically.

Author details

Kaida Xiao[1*], Sophie Wuerger[1], Faraedon Mostafa[3], Ali Sohaib[2] and Julian M Yates[2]

*Address all correspondence to: kaidaxiao@yahoo.co.uk

1 Department of Psychological Science, University of Liverpool, Liverpool, UK

2 Department of Oral and Maxillofacial Surgery, School of Dentistry, University of Manchester, Manchester, UK

3 Department of Periodontology, School of Dentistry, University of Sulaimani, Slemani, Iraq

References

[1] Worthington P, Branemark P, Advanced osseointegration surgery: applications in the maxillofacial region. Chicago: Quintessence, 1992.

[2] Khan Z, Gettleman L, Jacobson CS, Conference report: materials research in maxillofacial prosthetics. J Dent Res. 1992;71:1541–2.

[3] Leonardi A, Buonaccorsi S, Pellacchia V, Moricca LM, Indrizzi E, Fini G, Maxillofacial prosthetic rehabilitation using extraoral implants. J Craniofac Surg. 2008;19:398–405.

[4] Market JC, Lemon JC, Extraoral maxillofacial prosthetic rehabilitation at the M. D. Anderson Cancer Center: a survey of patient attitudes and opinions. J Prosthet Dent. 2001;85:608–13.

[5] Sachs E, Vezzetti E, Numerical simulation of deposition process for a new 3DP printhead design. J Mater Process Technol. 2005;161(3):509–515.

[6] Curodeau A, Sachs E, Caldarise S, Design and fabrication of cast orthopedic implants with freeform surface textures from 3-D printed ceramic shell. J Biomed Mater Res. 2000;53(5):525–535.

[7] Stopp S, Wolff T, Irlinger F, Lueth T, A new method for printer calibration and contour accuracy manufacturing with 3D-print technology. Rapid Prototyp J. 2008;14(3):167–1726.

[8] van Noort R. The future of dental devices is digital. Den Mater. 2011;28:3–12.

[9] Mavili ME, Canter HI, Saglam-Aydinatay B, Kamaci S, Kocadereli I, Use of three-dimensional medical modeling methods for precise planning of orthognathic surgery. J Craniofac Surg. 2007;18(4):740–7.

[10] Dimitrov D, Schreve K, de Beer N, Advances in three dimensional printing–state of the art and future perspectives. Rapid Prototyp J. 2006;12(3):136–147.

[11] Brunton A, Arikan CA, Urban P, Pushing the limits of 3D color printing: error diffusion with translucent materials. ACM Trans Graph. 2015;35:1–13.

[12] Schelly C, Anzalone G, Wijnen B, Pearce JM, Open-source 3-D printing technologies for education: bringing additive manufacturing to the classroom. J Visual Lang Comput. 2015;28:226–237.

[13] Aldridge K, Boyadjiev SA, Capone GT, DeLeon VB, Richtsmeier JT, Precision and error of three-dimensional phenotypicmeasures acquired from 3dMD photogrammetric images. Am J Med Genet. 2005;138A:247–253.

[14] Plooij JM, Swennen GRJ, Rangel FA, Maal TJJ, Schutyser FAC, Bronkhorst FM, Kuijpers-Jagtman AM, Berge SJ, Evaluation of reproducibility and reliability of 3D soft tissue analysis using 3D stereophotogrammetry. Int J Oral Maxillofac Surg. 2009;38:267–273.

[15] Stanic M, Lozo B, Svetec DG, Colorimetric properties and stability of 3D prints. Rapid Prototyp J. 2012;18(2):120–128.

[16] Xiao K, Zardawi F, van Noort R, Yates JM, Color reproduction for advanced manufacture of soft tissue prostheses. J Dent. 2013;41(S5):e15–e23.

[17] Xiao K, Zardawi F, van Noort R, Yates JM, Developing a 3D colour image reproduction system for additive manufacturing of facial prostheses. Int J Adv Manuf Technol. 2014;70:2043–2049.

[18] Zardawi F, Xiao K, van Noort R, Yates JM, Mechanical properties of 3D printed facial prostheses compared to handmade silicone polymer prostheses. Eur J Sci Res. 2015;11:12.

[19] Fairchild MD, Color Appearance Models. 3rd edition. Wiley. 2013.

[20] Gerhardt J, Hardeberg JY, Spectral color reproduction minimizing spectral and perceptual color differences. Color Res Appl. 2008;33:494–504.

[21] Ciocca L, Scotti R, CAD-CAM generated ear cast by means of a laser scanner and rapid prototyping machine. J Prosthet Dent. 2004;92:591–595.

[22] He Y, Xue G, Fu J, Fabrication of low cost soft tissue prostheses with the desktop 3D printer. Sci Rep. 2014;4:6973.

[23] Holberg C, Schwenzer K, Mahaini L, RudzkiJanson I, Accuracy of facial plaster casts. Angle Orthod. 2006;76:605–611.

[24] Liacouras P, Garnes K, Roman N, Petrich A, Grant GT, Designing and manufacturing an auricular prosthesis using computed tomography, 3-dimensional photographic imaging, and additive manufacturing: a clinical report. J Prosthet Dent. 2011;105:78–82.

[25] CIE Publ 159:2004. A Color Appearance Model for Color Management Systems. Vienna, Austria: CIE Central Bureau, 2004.

[26] ISO 15076-1:2005. Image technology color management—Architecture, profile format and data structure—Part 1: Based on ICC.1:2004–10.

[27] CIE S 014-3/E:2011 (ISO 11664-3:2012), Colorimetry—Part 3: CIE Tristimulus Values, STANDARD by Commission Internationale de L'Eclairage, 05/01/2012.

[28] Kaida Xiao, Julian M Yates, Faraedon Zardawi, Suchitra Sueeprasan, Ningfang Liao, Liz Gill, Changjun Li, Sophie Wuerger (2016) Characterising the variations in ethnic skin colours: a new calibrated data base for human skin. Skin Research and Technology (in press)

[29] Hong G, Luo MR, Rhodes PA, A study of digital camera colorimetric characterization based on polynomial modelling. Color Res Appl. 2001;26(1):76–84.

[30] Xiao K, Liao N, Zardawi F, Liu H, Noort RV, Yang Z, Huang M, Yates JM, An investigation of Chinese skin colour and appearance for skin colour reproduction. Chin Opt Lett. 2012;10:083301 1–5.

[31] CIE S 014-4/E:2007 (ISO 11664-4:2008), Colorimetry—Part 4: CIE 1976 L*a*b* Colour Spaces, STANDARD by Commission Internationale de L'Eclairage, 12/01/2008.

[32] Paravina RD, Majkic G, Del Mar Perez M, Kiat-amnuay S, Color difference thresholds of maxillofacial skin replications. J Prosthodont 2009;18:618–625.

[33] MIT (1989–2000) What is the 3DPTM process? Retrieved from the website of the 3DPTM Laboratory, Massachusetts Institute of Technology. Retrieved from: *http://web.mit.edu/tdp/www/whatis3dp.html. 11 December 2008.*

[34] Zardawi F, Xiao K, van Noort R, Yates JM, Investigation of elastomer infiltration into 3D printed facial soft tissue prostheses. Anaplastology 2015;4:139.

Chapter 5

3D-Printed Models Applied in Medical Research Studies

Jorge Roberto Lopes dos Santos, Heron Werner,
Bruno Alvares de Azevedo, Luiz Lanziotti,
Elyzabeth Avvad Portari, Sidnei Paciornik and
Haimon Diniz Lopes Alves

Additional information is available at the end of the chapter

http://dx.doi.org/10.5772/63942

Abstract

The aim of this chapter is to show experiments in cardiology and fetal medicine, two specialties of medicine, through the development of three dimensional (3D) physical models produced on additive manufacturing (AM) technologies, also known as 3D printing, from files acquired on noninvasive-imaging technologies (NITs) as 3D ultrasound (3DUS), magnetic resonance imaging (MRI), computed tomography (CT), and micro-computed tomography (micro-CT). The presentation of eight different experiments demonstrates that the combination of AM technologies and files obtained from NITs may improve our understanding of medical anatomical characteristics for medical research, simulation procedures, and educational purposes.

Keywords: additive manufacturing, 3D-printed models, 3D ultrasound, magnetic resonance imaging, computed tomography, cardiology, fetal medicine

1. Introduction

In recent years, the use of three-dimensional (3D) files generated in noninvasive-imaging technologies (NITs) combined with additive manufacturing (AM) technologies has increased exponentially. NITs are adopted to assist diagnosis in medicine allowing the physician to visualize and define relevant internal structures of the human body, as well as the conversion of the visualized information into a set of digital medical 3D data.

Eight selected experiments developed by a multidisciplinary team of physicians, designers, and physicists are described in order to highlight the potential of the combination of AM and NITs, demonstrating that 3D representation enhances the understanding for the diagnosis and treatment in medicine.

Four NITs were adopted so as to generate 3D imaging files: 3D ultrasound (3DUS), magnetic resonance imaging (MRI), computerized tomography (CT), and micro-computed tomography (micro-CT). The routine of imaging acquisition and the AM systems are similar regarding their logical method based on the virtual "slicing" of an object, generating an amount of layers according to the thickness recommended [1].

The consecutive construction of the projected 3D computer-aided design (CAD) model can be reached throughout the superimposition of those same layers. The additive materialization process begins with the 3D CAD model, which is then "sliced" into layers that contain spatial references to guide later the deposit of selected materials, layer by layer, resulting on a physical 3D model.

The actual additive method is based upon the successive overlapping of thin layers of specific material substances, according to the appropriate technical method, and is carried out by transforming the 3D files into an STL (Standard Triangulation Language) extension, which consists basically of *X*, *Y*, and *Z* coordinates. Once the STL file is generated, the next step is the horizontal slicing of the whole 3D volumetric file using software appropriate to the specific hardware being used, and calculating the supporting structures when necessary. The building process starts with the sequential deposition of layers of material, the layer width ranging from microns to fractions of millimeters, depending on the technology chosen [1].

This process is then followed by a post-processing stage, an essential procedure in all current AM technologies, where the model has to be cleaned to remove the support material and/or residues used during the building process. In the case of some AM processes that work by using a laser beam to harden photosensitive materials, it is also necessary to position and expose the model inside an ultraviolet light camera in order to solidify the model completely.

There are presently diverse systems of AM/3D printing technologies commercially available. Although they use different material processes, they are all based on the principle of physical materialization by layer deposition. One of the most important characteristics of the additive manufacturing technology is its capability of construction parts with any geometrical intricacy, a process in which subtractive technologies as CNC equipment are sometimes limited and time consuming [1–3].

2. 3D models in cardiology research projects

The use of imaging techniques in cardiology had its beginning with the advent of chest X-ray, in which a topographic image of the heart could be done in different projections (1). In this type of image, volumetric structures appeared overlapping each other in the X-ray film. Multiple projections helped the analysis of cardiac structures and the location of abnormalities. Another radioactivity utility was the use of radiotracers and detection chambers—developing the planar Gamma Cameras, in which two-dimensional images were generated by detecting gamma radiation emitted by the radiotracer distributed through the heart [4].

The idea of generating images as sequences slices of the heart gave rise to tomographic techniques. In this field, computing tomography was developed with the use of radiation generated by an electrical current: spect scintigraphy imaging, using the radiotracers, and magnetic resonance imaging, by acquiring energy emitted by accommodation of the nuclear spins. The possibility of three-dimensional reconstruction using two-dimensional tomographic images provided remarkable progress in the study of heart anatomy and diseases [5].

By the same principle of using two-dimensional images for three-dimensional reconstruction, the 3D echocardiography was developed. With this technique, it is possible to present the heart images in three dimensions. In addition, the heart movement is displayed, adding important information about the heart function, once it is a dynamic organ with constant movement [6–8].

In assessing the evolution of image acquisition techniques, topographic, two-dimensional slices, three-dimensional reconstruction, 3D moving images, we can see that fewer mental process and imagination is required to design an imaginary volumetric structure through their 2D projections. This fact seems to be important, because each individual has different abilities to mental reconstruction of a 3D structure. Such variability can bring a discrepancy in the evaluation of the same case by different physicians. For example, during surgical planning, there is no guarantee that the tomographic images analysis, by the surgeon and the radiologist, generated similar mental conceptions in the mind of each of these professionals [1, 9, 10].

As a next step, the construction of patient-specific 3D heart models, using rapid prototyping techniques, probably, will bring benefits to the cardiovascular science. Recent publications have pointed to the possibilities generated by this technology in cardiology. In education and medical training, 3D printing of heart structures can assist in improving the learning curve. On the diagnostic process, probably, the congenital and valvular heart disease may be better studied with the printing of physical models. In the treatment, especially in surgical interventions, patient-specific three-dimensional models can help as a pre-step procedure for a more assertive surgical planning. Perhaps, it can be said that the three-dimensional physical models, in a certain way, standardize the imaginary mental process of 3D reconstruction. Furthermore, in these models, the structures are displayed in a three-dimensional space, unlike the conventional 3D images displayed on a two-dimensional screen. The 3D environment is also present in holographic projections and virtual reality simulations, where the structures are manipulated and analyzed in a 3D virtual space. Although promising, the calibration and quality control of the equipment that are involved in the rapid prototyping technique are key factors. In each process step, image acquisition, digital file segmentation, 3D printing, and systematic errors can be added, resulting in a physical model that do not represent the real organ [10, 11].

Figure 1 shows a physical model, of a healthy heart of a medium height man, 50 years old. The digital archive of heart images was generated by performing an angio-CT scan, Somatom Sensation 64 × 0.6 mm (Siemens Inc., Germany).

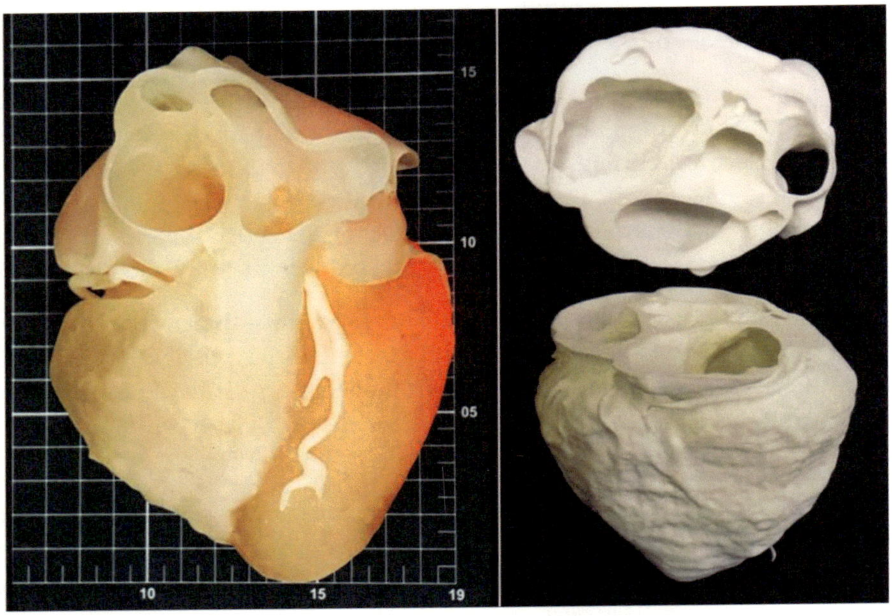

Figure 1. A. 3D model of human heart from CT scanner with "arduino" electronic board controlling led inside the model to allow the visualization of different densities of the heart positioned on measuring scale table (scale reference: 1 cm^2)—(3D printed in photosensible resin—PROJET 3510 HD Plus—3D Systems)—Núcleo de Experimentação Tridimensional—NEXT PUC-Rio; B: Same 3D file printed in two parts to allow the visualization of the internal structures (3D printed in polyamide at EOS P110)—Instituto Nacional de Tecnologia—Laboratório de Modelos Tridimensionais—Ministério da Ciência, Tecnologia e Inovação.

Aortic valve can be affected by many diseases that culminate with the necessity of a heart surgery, for aortic valve replacement. In this surgical procedure, the native aortic valve, dysfunctional, is excised and a biological or metallic prosthesis is implanted in the aortic annulus. After this surgery, several complications may occur. When the suture fixing the prosthetic valve in the aortic annulus breaks, a hole is formed between the prosthetic ring and the aortic annulus, a process named as paraprosthetic leak. With this complication, during ventricular diastole, blood present in the aortic root returns to the left ventricle, causing volume overload that can result with a severe left ventricular dysfunction.

Figure 2A presents a physical model, in full scale, of a heart segment in which there is a paraprosthetic leak around a biological aortic valve prosthesis. This hole measures approximately 1 cm. This model was carried out with the purpose of planning the percutaneous procedure that would be done to fix this abnormality. In **Figure 2B** and **C**, the implantation process of an occluder may be noted within the paraprosthetic leak. In this case, the physical model can help the physician to detect the spatial location of the defect to be corrected before the procedure is started. Besides that, the physical model can be used to test many types and sizes of percutaneous occluders. The digital archive of heart images was generated by performing an angio-CT scan, Somatom Sensation 64 × 0.6 mm (Siemens Inc., Germany).

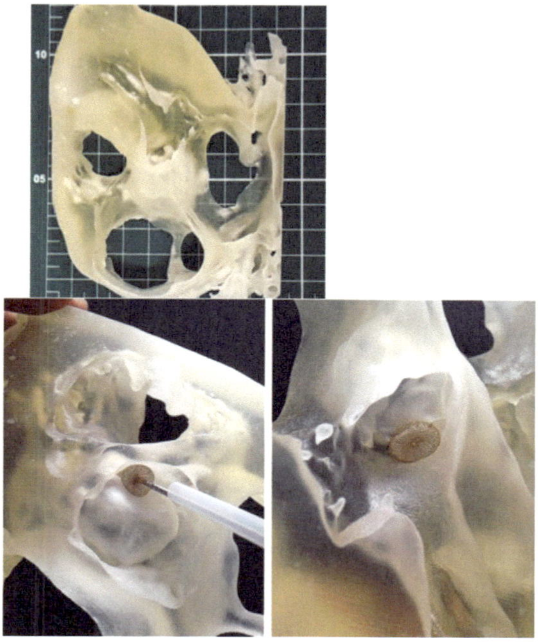

Figure 2. A. 3D model of heart section from CT scanner files positioned on measuring scale table (scale reference: 1 cm^2); B and C: detail highlighting paravalvular leak beside aortic bioprosthesis (3D printed in photosensible resin—PROJET 3510 HD Plus - 3D Systems)—Núcleo de Experimentação Tridimensional—NEXT PUC-Rio.

3. Aortic aneurysms

In the field of vascular surgery, particularly in the case of aortic stent grafting of the treatment of aortic aneurysms, there is already a commercially available product in which the stent graft is custom-made for each patient with unusual complex anatomy relaying on the 3D printing technology. To provide the surgeon with training, he receives in advance a prototype of the custom-made stent graft and a 3D model of the patient's aorta. If needed, modifications can be suggested until there is agreement between the engineering and the medical team [12, 13].

The evolving technology allows for flexible and malleable models that are fairly realistic in its shape (**Figure 3A** and **B**). Another practical application already in use is medical training itself, in which the training vascular surgeon can learn the so-called endovascular techniques by deploying stents and stent grafts in 3D artery models (**Figure 3C** and **D**). These techniques are normally learned during hundreds of hours of "hands-on" training under staff supervision, which carry along, for both patients and surgeons, extra doses of radiation, required to perform these interventional radiology procedures. The current underdeveloping 3D training models do not require radiation and give the training surgeon a 3D hands-on feeling and vision of the devices itself, since we can print transparent flexible models, while the actual endovascular procedures provide only with the regular 2D images of the monitor screen (**Figure 4**).

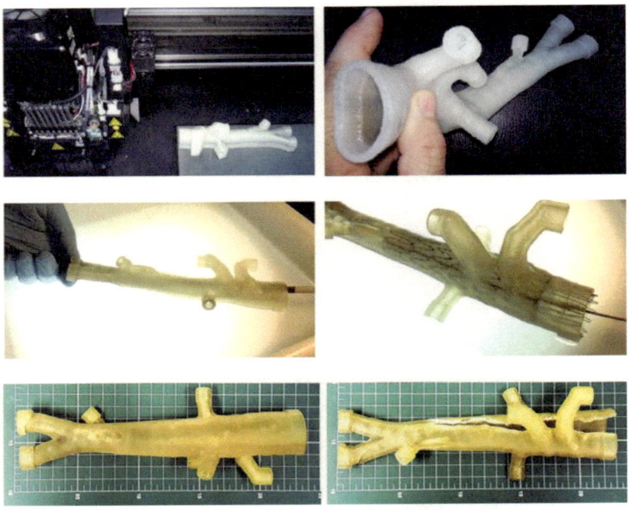

Figure 3. A and B. Printing process and the flexible 3D aorta model; C and D: deployment of an aortic stent graft into an aortic model, serving as educational resource for training vascular surgeons; E: model positioned on measuring scale table (scale reference: 1 cm^2); F: model damaged after tests. (3D printed in flexible clear resin on OBJET Connex 350)—Instituto Nacional de Tecnologia—Laboratório de Modelos Tridimensionais—Ministério da Ciência, Tecnologia e Inovação.

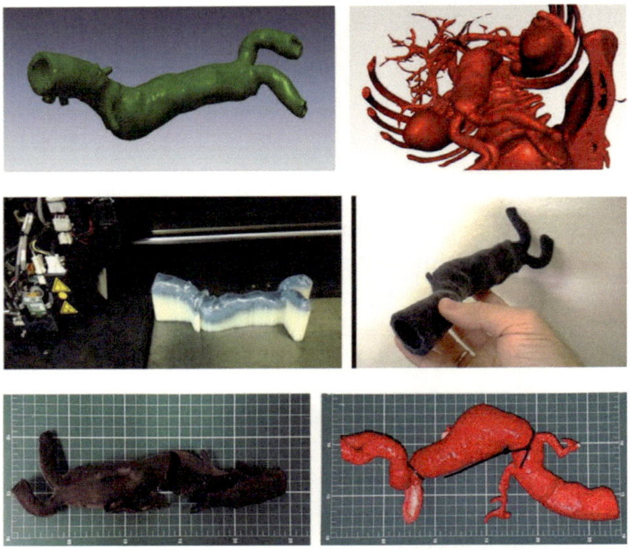

Figure 4. A–D. Sequence of the development of a 3D flexible model of aorta from CT scanner files; E: 3D printed flexible model of aorta positioned on measuring scale table (scale reference: 1 cm^2), damaged after tests (3D Printed in "Tango black" flexible resin on OBJET Connex 350)—Instituto Nacional de Tecnologia—Laboratório de Modelos Tridimensionais—Ministério da Ciência, Tecnologia e Inovação.; F: variant colored rigid 3D printed model of aorta (3D printed in thermoplastic ABS—Stratasys U-Print-Fused Deposition Modeling)—Núcleo de Experimentação Tridimensional—NEXT PUC-Rio.

4. Mitral regurgitation

The mitral valve prolapse, the well-known entity, in some specific cases may be associated with mitral regurgitation. When mitral regurgitation reaches a high level, a large volume of blood returns to the left atrium during ventricular systole. In these cases, there may be important symptoms culminating with left ventricular dysfunction. When this occurs, a mitral regurgitation correction procedure is indicated. One surgical technique, already established for this condition, is the Quadrangular Resection. When a severe mitral regurgitation is caused by prolapse of the posterior leaflet, the Quadrangular Resection can be performed. In this surgical technique, part of valve leaflet tissue is excised, besides the confection of a suture in the mitral annulus. The objective of this procedure is to enable the full coaptation of the leaflets during ventricular systole, preventing blood regurgitation into the left atrium [7].

Figure 5A presents, in three projections, the mitral valve leaflets of a patient with prolapse of the posterior leaflet, which cause no coaptation and a regurgitant orifice. Again, as a step to plan the surgical procedure, a virtual simulation of Quadrangular Resection was performed in the digital file using Rhinoceros 5.0 (McNeel North America, USA). In this simulation, the amount of tissue that would be needed to be excised from the posterior mitral leaflet can be estimated (**Figure 5B**).

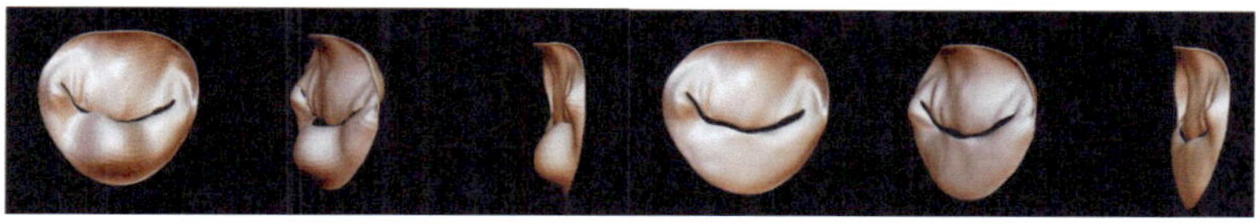

Figure 5. 3D virtual models of Quadrangular Resection in mitral valve repair—Núcleo de Experimentação Tridimensional—NEXT PUC-Rio.

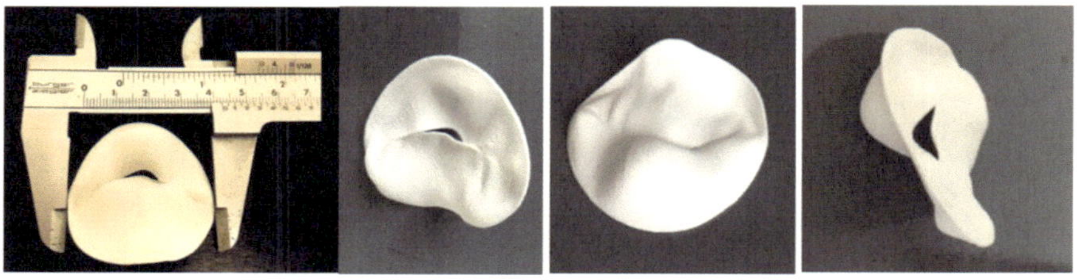

Figure 6. A–D. 3D printed model of mitral valve prolapse from 3D ultrasound (3D US)—printed in poliamide—(EOS P110)—Instituto Nacional de Tecnologia—Laboratório de Modelos Tridimensionais—Ministério da Ciência, Tecnologia e Inovação.

This file is also used for making the physical model, which can be seen from different angles in **Figure 6A–D**. The digital file of the mitral valve leaflets was generated by performing a 3D Echocardiography with Vivid E9 (GE Healthcare, USA).

5. Fetal medicine

Advances in image-scanning technology have led to vast improvements in fetal medicine [14–16].

In general, three main technologies are used to obtain images within the uterus during pregnancy—ultrasound, MRI, and CT. US is currently the primary method for fetal assessment during pregnancy because it is patient-friendly, useful, cost-effective, and considered to be safe. When ultrasound yields equivocal results, MRI can be used to help in diagnosis. CT can be used after 30 weeks of gestation only in cases of skeletal dysplasias [3, 17].

The development of ultrasound scanning during the 1960s opened a new window into the study of the fetus. Its introduction in the clinical obstetrics practice allowed the opportunity to make screening of many malformations (**Figures 7–10**).

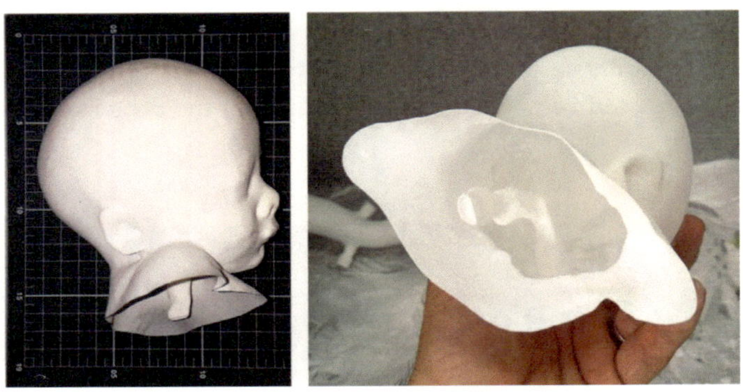

Figure 7. A and B. 3D printed hollow model of a head of a 32-week fetus from magnetic resonance imaging (MRI) highlighting fetal upper respiratory tract model positioned on measuring scale table (scale reference: 1 cm^2)—3D printed in polyamide (EOS P110)—Instituto Nacional de Tecnologia—Laboratório de Modelos Tridimensionais—Ministério da Ciência, Tecnologia e Inovação.

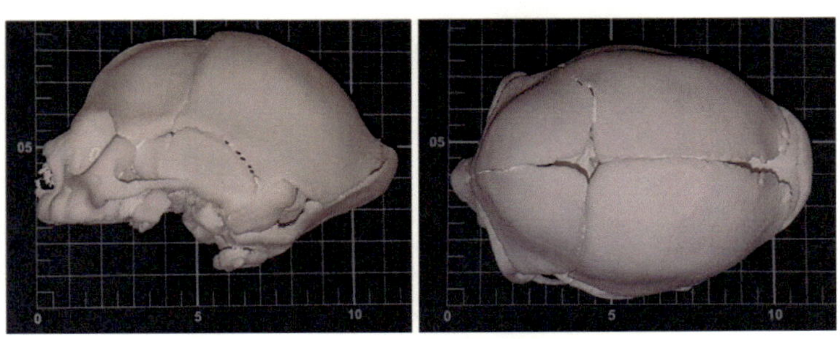

Figure 8. A and B. 3D full-scale model of the skull of neonate with microcephaly from CT scanner, model positioned on measuring scale table (scale reference: 1 cm^2)—3D printed in polyamide (EOS P110)—Instituto Nacional de Tecnologia—Laboratório de Modelos Tridimensionais—Ministério da Ciência, Tecnologia e Inovação.

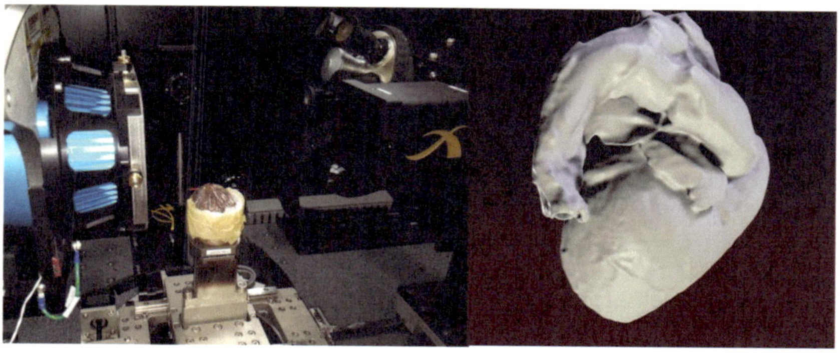

Figure 9. A and B. Fetal heart at 38 weeks, after autopsy, being positioned inside a micro-CT scanner (Zeiss Xradia Versa 510) and 3D virtual model generated from the DICOM micro-CT files—Department of Chemical and Materials Engineering—Rio de Janeiro, Brazil—DEQM/PUC-Rio and 3D printed models (3D Systems PROJET 3510 HD Plus) — Núcleo de Experimentação Tridimensional—NEXT PUC-Rio.

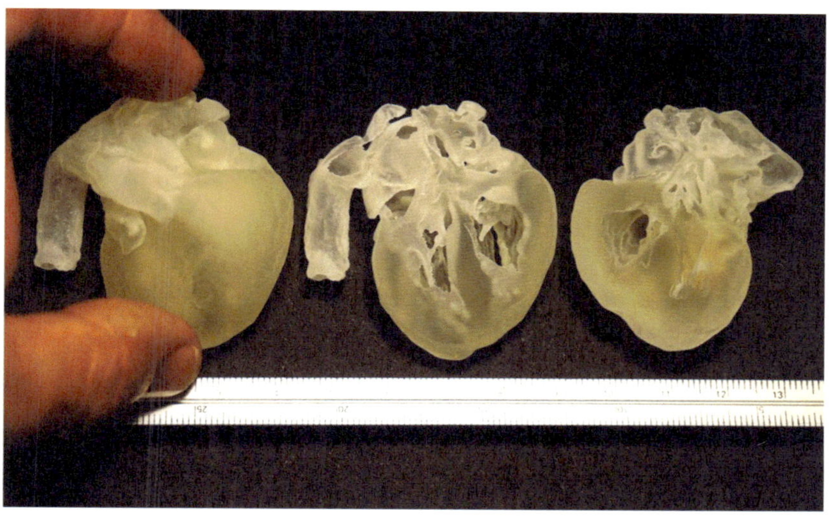

Figure 10. 3D printed full-scale model of human fetal heart (38 weeks) from micro-CT scanner files —SLA system (3D Systems PRO-7 JET 3510 HD Plus)—Núcleo de Experimentação Tridimensional—NEXT PUC-Rio.

MRI is a noninvasive procedure that has been utilized in obstetrics since the 1980s. It offers digital files with exceptional image contrast allowing a great visual comprehension of the fetus internal tissues. When the use of ultrasound detects or produces unexpected results, MRI is often adopted, since it provides further information about fetal malformations and conditions for which ultrasound cannot generate superior images [18].

MRI files can generate detailed characteristics of the soft tissues of the fetus body as the face, hands, or feet as well as internal body structures as aerial paths. The construction process of the 3D accurate virtual model starts with the 3D modeling volume built through the MRI slices sequentially mounted, followed by the segmentation process where the physician selects the important body parts to be analyzed that will then be reconstructed in 3D [19].

Fetal 3D physical models from 3DUS and MRI scan data have been used in blind pregnant women to improve maternal fetal attachment [20, 21].

6. Conclusion

3D printing has become a very useful and potentially transformative tool in a number of different fields, including medicine. It is an emerging technique with a variety of medical applications such as surgical planning and training, implant design, biomedical research, and medical education. There is an enormous potential of the technique and in the near future growing utilization and development of new applications in the fields of individual patient care, as well as academic and research activities will occur.

Researchers continue to improve existing medical applications that use 3D printing technology and to explore new ones. 3D printing provides a new didactic approach to learning when applied to medicine. The opportunity to prospect this paradigm shift in knowledge that occurs in the transition from 2D to 3D enables doctors to learn more inductively through the 3D modeling as a learning tool and improve patient's care.

Author details

Jorge Roberto Lopes dos Santos[1,2], Heron Werner[3*], Bruno Alvares de Azevedo[4], Luiz Lanziotti[5], Elyzabeth Avvad Portari[6,7], Sidnei Paciornik[8,9] and Haimon Diniz Lopes Alves[10,11]

*Address all correspondence to: heronwerner@hotmail.com

1 Pontifícia Universidade Católica do Rio de Janeiro, Arts and Design Department, Núcleo de Experimentação Tridimensional, Rio de Janeiro, Brazil

2 Ministério da Ciência, Tecnologia e Inovação, Instituto Nacional de Tecnologia, Laboratório de Modelos Tridimensionais, Rio de Janeiro, Brazil

3 Clínica de Diagnóstico por Imagem, CDPI, Rio de Janeiro, Brazil

4 Pontifícia Universidade Católica do Rio de Janeiro, Department of Mechanical Engineering, Rio de Janeiro, Brazil

5 SBACV, Brazilian Society of Angiology and Vascular Surgery, Rio de Janeiro, Brazil

6 Universidade Estadual do Rio de Janeiro, Rio de Janeiro, Brazil

7 Fundação Oswaldo Cruz, Instituto Nacional de Saúde da Mulher, da Criança e do Adolescente Fernandes Figueira, Rio de Janeiro, Brazil

8 Pontifícia Universidade Católica do Rio de Janeiro, Department of Chemical and Materials Engineering, Rio de Janeiro, Brazil

9 DEQM/PUC-Rio, Rio de Janeiro, Brazil

10 Universidade Estadual do Rio de Janeiro, Rio de Janeiro, Brazil

11 Pontifícia Universidade Católica do Rio de Janeiro, Department of Chemical and Materials Engineering, Rio de Janeiro, Brazil

References

[1] Santos JL, Werner H, Fontes R, et al. Additive manufactured models of fetuses built from 3D ultrasound, magnetic resonance imaging and computed tomography scan data. In: Muhammad Enamul Hoque (Ed), Rapid Prototyping Technology – Principles and Functional Requirements. InTech Rijeka, Croatia, 2011, 179–192.

[2] Fiorini ST, Frajhof L, Azevedo BA, Santos JRL, Werner H, Raposo A, de Lucena CJP. Three-dimensional models and simulation tools enabling interaction and immersion in medical education. Des User Exp Usability. 2015; 9188: 662–671.

[3] Werner H, Dos Santos JRL, Fontes R, Daltro P, Gasparetto E, Marchiori E, Campbell S. Additive manufacturing models of fetuses built from three-dimensional ultrasound, magnetic resonance imaging and computed tomography scan data. Ultrasound Obstet Gynecol. 2010; 36: 355–361.

[4] Lombardi CM, Zambelli V, Botta G, Moltrasio F, Cattoretti G, Lucchini V, Fesslova V, Cuttin MS. Postmortem microcomputed tomography (micro-CT) of small fetuses and hearts. Ultrasound Obstet Gynecol 2014; 44: 600–609.

[5] Matthew D.B.S. Tam, Stephen D. Laycock, James R.I. Brown, and Matthew Jakeways. 3D Printing of an aortic aneurysm to facilitate decision making and device selection for endovascular aneurysm repair in complex neck anatomy. J Endovasc Ther. 2013; 20: 863–867. doi:10.1583/13-4450MR.1

[6] Farooqi KM, Sengupta PP. Echocardiography and three-dimensional printing: sound ideas to touch a heart. J Am Soc Echocardiogr 2015; 28: 398–403.

[7] Houck RC, Cooke JE, Gill1 EA. Live 3D echocardiography: a replacement for traditional 2D echocardiography? Am J Roentgenol 2006; 187: 1092–1106.

[8] Madsen EB, Gilpin E, Slutsky RA, Ahnve S, Henning H, Ross Jr. J. Usefulness of the chest x-ray for predicting abnormal left ventricular function after acute myocardial infarction. Am Heart J 1984; 108: 1431–1436.

[9] Armillotta A, Bonhoeffer P, Dubini G, et al. Use of rapid prototyping models in the planning of percutaneous pulmonary valve stent implantation. Proc Inst Mech Eng H 2007; 221: 407–416.

[10] Gillinov AM, Cosgrove DM. Mitral valve repair for degenerative disease. J Heart Valve Dis 2002; 11: 515–520.

[11] Heo J, Cave V, Wasserleben V, Iskandrian AS. Planar and tomographic imaging with technetium 99m-labeled tetrofosmin: Correlation with thallium 201 and coronary angiography J Nucl Cardiol 1994; 1: 317–324.

[12] Mäkelä T, Clarysse P, Sipilä O, Pauna N, Pham Q, Katila T, Magnin I. A review of cardiac image registration methods. IEEE Trans Med Imaging 2002; 9: 1011–1021.

[13] Matsui H, Ho SY, Mohun TJ, Gardiner HM. Postmortem high-resolution episcopic microscopy (HREM) of small human fetal hearts. Ultrasound Obstet Gynecol 2015; 45: 492–493.

[14] Werner H, Lopes Dos Santos JR, Araújo Junior E. Three-dimensional virtual sonographic and magnetic resonance imaging in obstetrics. Adv Top Three-dimensional Ultrasound ObstetGynecol 2016; 1: 227–238.

[15] Werner H, Lopes dos Santos J, Araújo Junior E. Physical models of the foetus created using magnetic resonance imaging, computed tomography, and ultrasound data: history, description, and potential uses. Rev Bras Ginecol Obstet 2015; 37(4): 149–151.

[16] Werner H, Dos Santos JRL, Fontes R, et al. Virtual bronchoscopy for evaluating cervical tumors of the fetus. Ultrasound Obstet Gynecol 2013; 41: 90–94.

[17] Werner H, Dos Santos JR, Fontes R, et al. The use of rapid prototyping didactic models in the study of fetal malformations. Ultrasound Obstet Gynecol 2008; 32: 955–956.

[18] Werner H, Rolo LC, Junior EA, Lopes Dos Santos JR. Manufacturing models of fetal malformations built from 3-dimensional ultrasound, magnetic resonance imaging, and computed tomography scan data. Ultrasound Q 2014; 30: 69–75.

[19] Werner H, Lopes J, Tonni G, Junior EA. Physical model from 3D ultrasound and magnetic resonance imaging scan data reconstruction of lumbosacral myelomeningocele in a fetus with Chiari II malformation. Childs Nerv Syst 2015; 31: 511.

[20] Campbell S. 4D and prenatal bonding: still more questions than answers. Ultrasound Obstet Gynecol 2006; 27: 243–244.

[21] Werner H, Lopes J, Tonni G, Araujo Júnior E. Maternal-fetal attachment in blind women using physical model from three-dimensional ultrasound and magnetic resonance scan data: six serious cases. J Matern Fetal Neonatal Med 2015; 1: 1–4.

Chapter 6

3D Printing Cardiovascular Anatomy: A Single-Centre Experience

Giovanni Biglino, Claudio Capelli,
Andrew M. Taylor and Silvia Schievano

Additional information is available at the end of the chapter

http://dx.doi.org/10.5772/63411

Abstract

This chapter presents the experience of the cardiac engineering team within the Centre for Cardiovascular Imaging at Great Ormond Street Hospital for Children (London, UK) in using three-dimensional (3D) printing technology. 3D models can serve different functions towards implementing a patient-specific approach for studying and potentially treating congenital heart disease (CHD). In order to showcase different potential applications, this chapter discusses not only clinical case studies and engineering experiments but also the potential for translation through patients and public involvement and engagement (PPI/E).

Keywords: 3D printing, additive manufacturing, congenital heart disease, intervention, patient-doctor communication

1. Introduction

1.1. 3D printing in medicine and congenital heart disease

3D printing technology represents an extraordinary innovation, which has been compared to a 'second industrial revolution' [1]. In May 2015, an article in the Harvard Business Review commented that 'industrial 3-D printing is at a tipping point, about to go mainstream in a big way. Most executives and many engineers don't realise it, but this technology has moved well beyond prototyping, rapid tooling, trinkets, and toys' [2]. Indeed, applications of 3D printing technology now range from the automotive industry to archaeology, from eyewear and fashion design to medicine. It is the medical field that experts in the sector indicate as the area in which the technology will play a really substantial and potentially revolutionary role in the coming years [3]. But 3D printing has been used for medical purposes well before recent technological advances and the press hype that ensued, with highly sensational titles in the media along the lines of '3D printer saves baby's life' [4]. Examples of medically relevant 3D models date back to the 1990s. Areas of medicine in which 3D printing technology can play an important role include manufacturing maxillofacial implants [5], personalising orthopaedic prostheses [6], creating patient-specific implants for craniofacial surgery [7], and even reconstruction of wound models for forensic purposes [8]. From renal [9] to cardiovascular [10], a range of models has now been manufactured and described, based on the underpinning principle that 3D printing technologies allow to print patient-specific models reconstructed from volumetric medical imaging data, based on a processing framework discussed in more detail elsewhere [11] and summarised in **Figure 1**.

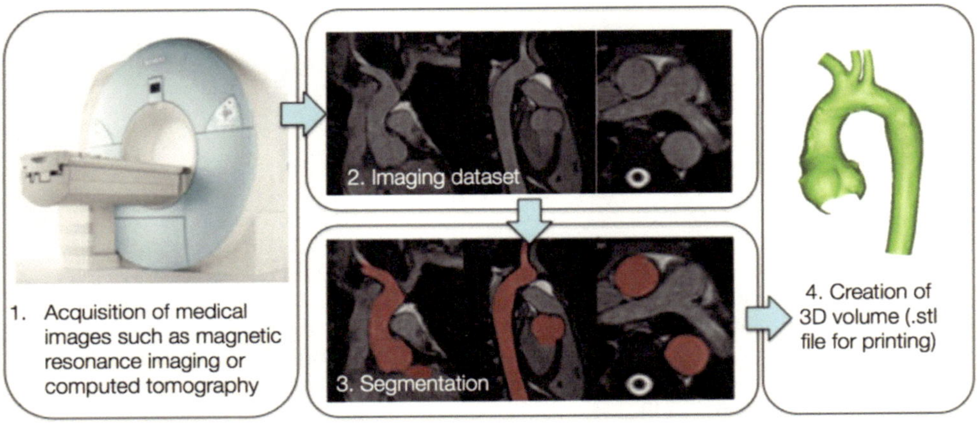

Figure 1. Framework for 3D printing cardiovascular 3D models, starting from medical imaging acquisition (1), processing the image dataset with appropriate software (2 and 3) and generating a 3D volume (4) in a format compatible with the 3D printer.

The principle of being able to reconstruct patient-specific anatomical models and manufacture them by means of 3D printing (additive manufacturing, in particular) is especially appealing in the field of congenital heart disease (CHD). Patients with CHD are in fact born with complex, often unique, vascular and/or intra-cardiac anatomical arrangements. Parts of these complex anatomies get further manipulated and modified following cardiac surgery, with vessels being transposed, baffles being created, narrowed (i.e. stenosed) vessels being enlarged and whole areas (e.g. right ventricular outflow tract) being augmented by patches. The same patient can also receive implantation of cardiac devices, such as stents, valves prosthesis or occluder devices, during catheterisation procedures. Overall, the resulting unique scenario certainly warrants a patient-specific approach, which can be beneficial for different purposes:

- Appreciation of anatomy for patients selection or device selection
- Experimentation for better appreciation of the physiology and haemodynamics

- Opportunity to practice new procedures and test new devices
- Design of novel training and communication tools

Based on this rationale, the cardiac engineering team within the Centre for Cardiovascular Imaging at Great Ormond Street Hospital for Children (London, UK) has advocated a patient-specific approach for studying and potentially treating CHD for several years now. In this light, 3D printing technology has been used extensively and for different purposes, focusing on the potential translation of the technique in the clinical realm. This chapter briefly presents an overview of the Centre's experience with 3D printing CHD, in order to create a showcase of the potential of the technique in the field. It will therefore touch upon clinical case studies, engineering experiments, and patients & public involvement and engagement (PPI/E).

2. Early experience: testing a novel cardiovascular device

One important application of 3D printing in CHD is represented by aiding in patient selection for specific procedures. In this context, the usefulness of 3D-printed models was assessed specifically for evaluating shapes and sizes of the right ventricular outflow tracts (RVOTs) in patients that would be possible candidates for percutaneous pulmonary valve implantation (PPVI). As a Centre leading in the introduction of PPVI in clinic [12] as well as its evaluation with engineering analyses [13], 3D printing patient-specific models represented an additional valuable tool for determining PPVI suitability. A study [14] demonstrated that having access to 3D models in addition to cardiovascular magnetic resonance (CMR) imaging data resulted in cardiologists improving the indication of patients suitable for implanting the first PPVI device (Melody® valve, Medtronic Inc) (**Figure 2**).

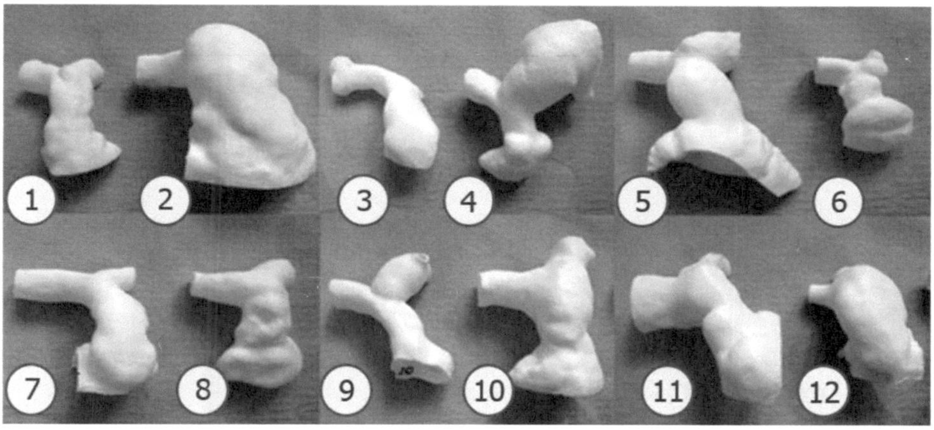

Figure 2. Example of 3D-printed RVOTs from 12 patients [ref].

Within the context of PPVI and in relation to early observation on the first generation device being suitable for only a small portion of patients requiring pulmonary valve replacement, a novel device was developed. The latter was designed with the intention of fitting more dilated RVOTs. In this case, a patient-specific 3D model resulted extremely valuable for practicing device insertion prior to performing the first implantation of the novel device on compassionate grounds (i.e. first-in-man procedure). A candidate (42-year-old male) presented with a history of cardiac surgeries, deemed not suitable for additional surgery, and, however, still presenting with severe pulmonary regurgitation. The morphology and dimensions of his RVOT suggested that the second-generation PPVI device would have been suitable. A model was created from his computed tomography (CT) examination and 3D printing. As shown in **Figure 3**, the cardiologist was then able to use the model to practice device insertion and deployment prior to the procedure. Particularly, having access to the model allowed the cardiologist a) to guide device personalisation to assure safe anchoring in the specific anatomy and b) to identify a problem in deploying the device if accessing via the right pulmonary artery —conventional approach—while he could successfully deploy it when inserting the line via the left pulmonary artery. The latter strategy was used (successfully) on the day of the procedure [15].

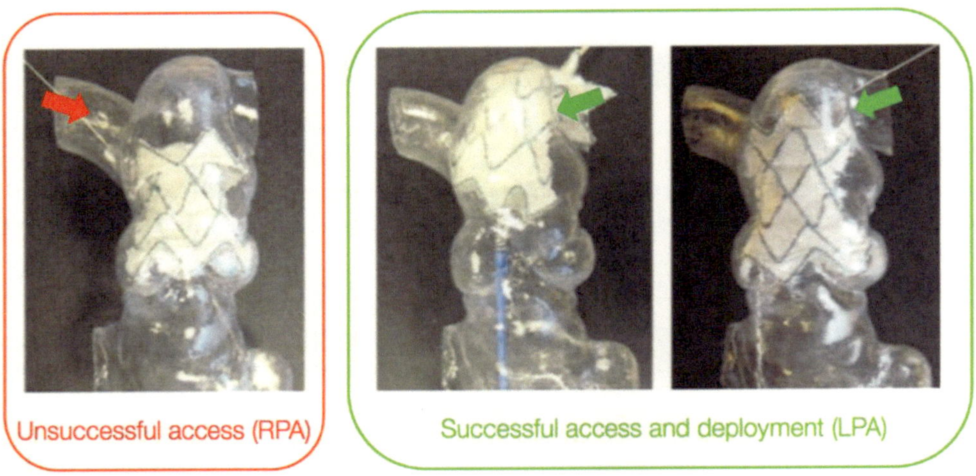

Figure 3. A 3D-printed model was used for testing insertion and deployment of a novel device (second-generation, self-expandable PPVI stent). As a valuable practice tool, testing the procedure in the model is allowed to identify an unfeasible access route. LPA, left pulmonary artery; RPA, right pulmonary artery. . Stent diameter at extremities = 40.7 mm.

3. Models for decision support: a case study

By establishing a conversation with clinicians and embedding the technology within the hospital, it is possible that complicated cases requiring additional insight are brought to the attention of the engineer for the creation of a model, which could be either a computer simulation or a 3D-printed model, or both. The model is created specifically to appreciate features that have clinical implications. An interesting example is represented by a patient (19-year-old male) who presented a diagnosis of aortic coarctation, previously stented, and also had an aberrant left subclavian artery originating from the stent site. The discussion revolved around the possibility of inserting a Cheatham Platinum (CP) covered stent within the existing device to prevent formation of a small developing aortic aneurysm, without occluding the aberrant subclavian artery. Insight into the relative position of all these components was therefore desirable, and a model was reconstructed from CT data, including the stent. The aorta was printed in transparent rigid resin, while the stent was printed using a white stereolithography (SLA), therefore being easily identifiable within the anatomy. Furthermore, a cut was performed alongside the thoracic/descending aorta such that the model could be easily opened and, if necessary, to gather additional measurements. The model (**Figure 4**) contributed to assess the feasibility of the intervention highlighting the spatial relationship between stent, aneurysm and left subclavian artery. Additional computational analyses (including the same 3D domain) were performed to simulate the implantation of the CP-covered stent and the local fluid dynamics post implantation [16]. The intervention was performed successfully.

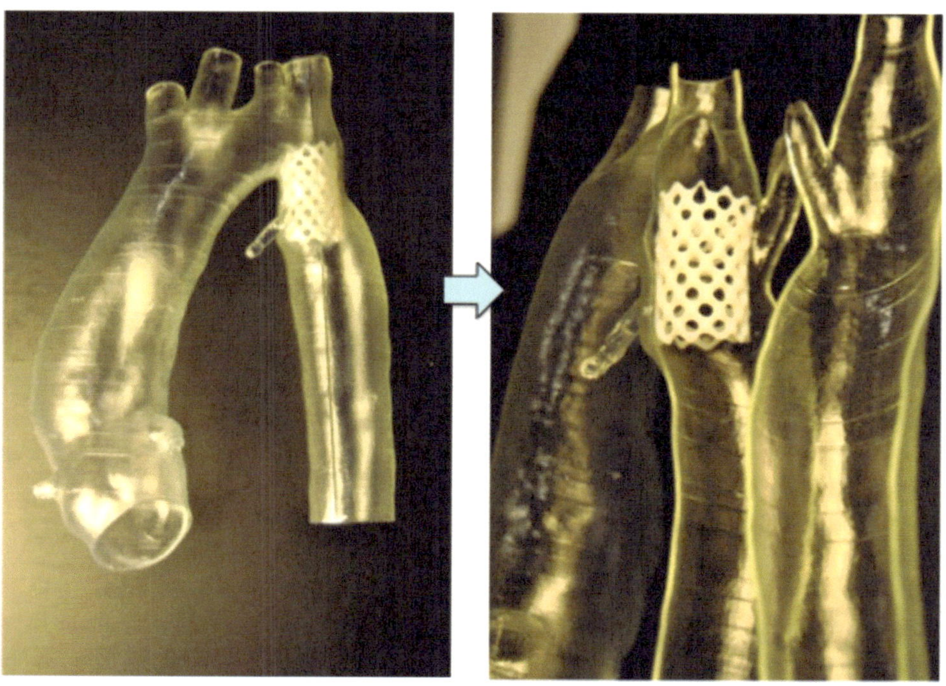

Figure 4. Patient-specific 3D model of stented coarctation, printed with two different materials to better highlight the position of the device (length = 28 mm), including the possibility of opening the model to better appreciate the relative position of the stent with respect to the rest of the anatomy.

4. Developing patient-specific models for research purposes

The possibility of printing 3D patient-specific CHD models is appealing not only for important clinical applications (e.g. testing devices and procedures) but also for research purposes. Models can in fact be incorporated into experimental setups for the detailed analysis of local fluid dynamics and for overall hydrodynamic measurements. This is of particular relevance in CHD patients, as often in vivo measurements are not feasible or, if feasible, are technically challenging, e.g. measuring pressure-volume loops in single ventricle patients [17]. The use of appropriately tuned experimental systems has the advantage that parametric studies can be performed in a controlled and reproducible setting, i.e. changing one variable at a time incrementally. This can generate hydrodynamic data which can increase the understanding of complex physiology.

4.1. Multi-scale hydrodynamic setups: studying single ventricle patients

Patients with single-ventricle physiology (e.g. hypoplastic left heart syndrome, HLHS) present with a fragile physiology, as their cardiovascular system relies on a single pumping chamber supporting both the systemic and pulmonary circulations [ref]. Hydrodynamic measurements are relevant to assess the effect of comorbities, e.g. aortic coarctation, on the fluid dynamics, as well as accurate measurements of quantities that are difficult to acquire in vivo, e.g. pressure drop in the surgically enlarged aorta in patients with palliated HLHS. To this end, a mock circulatory loop (MCL) can be constructed, whereby the cardiovascular system is experimentally summarised by appropriately tuned resistive and compliant elements. A system was built [18] using a Berlin Heart® ventricular assist device to simulate the single ventricle, and 3D-printed patient-specific models were manufactured and 'plugged' into the MCL to perform parametric experiments. This allowed us to test different aortic morphologies in the same controllable setup, e.g. tubular aortic arch, aortic arch with a severe coarctation, or an aortic arch in which the patching performed during the Norwood procedure had resulted in a very dilated transverse arch therefore creating a size mismatch with the descending aorta. 3D models are shown in **Figure 5**. Data on pressure drop along the aorta or flow distribution (upper body, lower body, pulmonary circulation) can be obtained using catheters and flow probes. MCLs can also be modified to simulate different surgical procedures. In the case of HLHS, for example, different surgical options exist to perform the first palliative procedure, i.e. different shunts can be used to source pulmonary blood flow to the pulmonary circulation. An MCL was built simulating first-stage palliation of HLHS with a modified Blalock-Taussig shunt [18] from the innominate artery to the pulmonary bed of the circuit and then modified to simulate first-stage palliation of HLHS with a Sano shunt [19] from the ventricle itself (in the experimental setup, from the de-airing valve of the Berlin Heart®) to the pulmonary bed of the circuit. The coupling of a detailed 3D model with a circuit summarising the remainder of the circulation can be considered a multi-scale [20] in vitro modelling approach for CHD. With regard to HLHS palliative surgery and the valuable insight that can be gathered experimentally, MCLs can be modified to incorporate the effect of respiration [21] which becomes relevant when these patients receive their second and third surgeries, i.e. the Glenn procedure and the total cavopulmonary connect (or Fontan completion) [22]. A system incorporating a patient-specific 3D-printed Fontan model has indeed been realised [23].

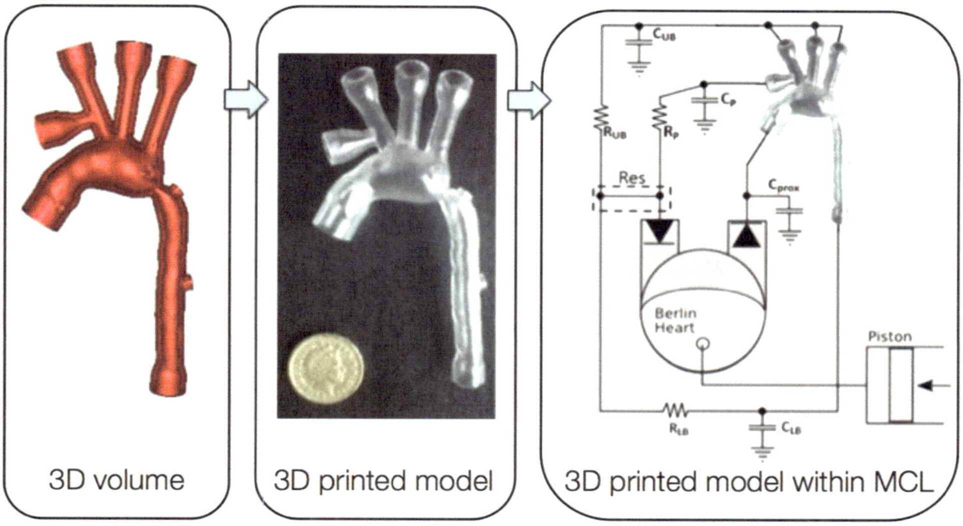

Figure 5. Model of the aortic arch of a baby with hypoplastic left heart syndrome following the Norwood operation and including a visible aortic coarctation. The 3D volume (left) is modified for the purpose of carrying out hydrodynamic experiments, hence including funnelled attachments to the MCL and ports for pressure measurements along the aorta. The volume is 3D printed with rigid resin (centre) and inserted in the MCL (right), which is here represented with its electrical equivalents, including a Berlin Heart® device to simulate the single ventricle.

4.2. Visualisation studies and validation of computational models

An MCL which incorporates 3D-printed anatomical models can be utilised to two additional ends: (a) serving the purpose of validation test beds for computational models and (b) generating detailed information on local fluid dynamics by means of visualisation experiments. In the first case, computational models have been introduced as a powerful resource for generating knowledge in CHD and possibly aiding in the decision-making process [24], but, in order for numerical models to be reliable, they require thorough validation against 'real world data' [25] which can be obtained in vitro. In the second case, the MCL can be modified to be compatible with different visualisation techniques, e.g. particle image velocimetry [23] or magnetic resonance imaging, thereby generating detailed visualisation data. A case study of CHD, namely transposition of the great arteries (TGA) repaired with the arterial switch procedure (ASO), was modelled for both purposes [26]. In this study, an MCL was built in order to be compatible with CMR image acquisition, and a 3D patient-specific model was inserted in the MCL for visualisation purposes. The model was reconstructed from CMR data of a patient (16-year-old male) with repaired TGA with ASO and Lecompte manoeuvre, showing typical features of TGA aortic arches (i.e. a dilated aortic root and a gothic aortic arch), thereby representing a representative test bed for studying haemodynamics in this clinical scenario. The model was 3D printed with transparent rigid resin. Visualisation data were acquired with 4D CMR flow, which generates exquisite images of flow streamlines within the volume of interest over the cardiac cycle. An age-matched control model (i.e. no CHD) was included in the study for comparison purposes. Furthermore, the model was replicated in silico in a multi-scale model including the same 3D domain (i.e. TGA or control aortic arch) and a lumped parameter network summarising the remainder of the circulation. Qualitative and quantitative comparison of flow and pressure data was carried out in order to validate the computational model, demonstrating good agreement between the two for both TGA and control scenarios. An example for this modelling paradigm is shown in **Figure 6**.

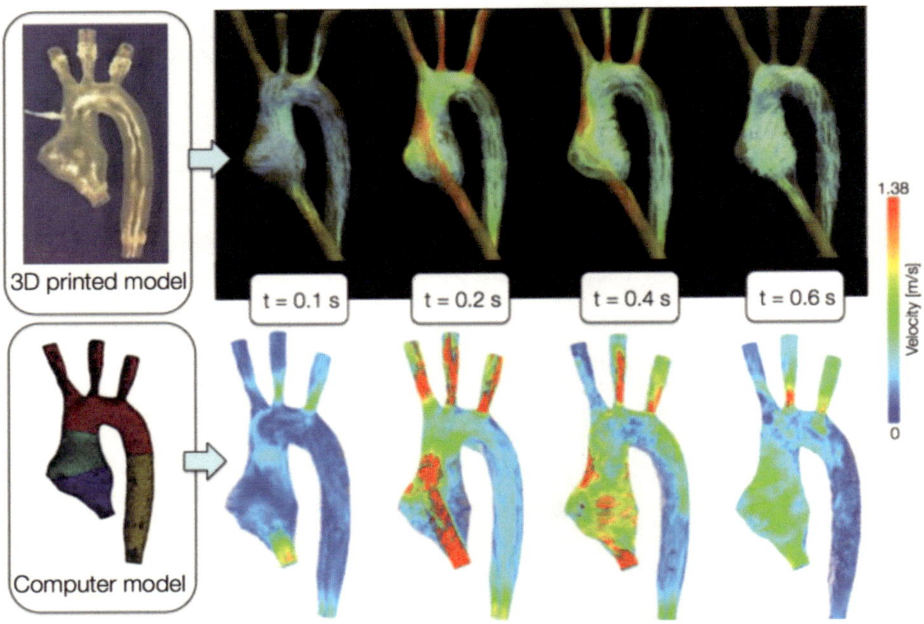

Figure 6. A model of the aortic arch 3D printed for visualization CMR experiments (top) and meshed for computational simulations (bottom). The images show example of flow-velocity streamlines from both CMR and computational fluid dynamics at the same time points through the cardiac cycle, showing good qualitative agreement as well as agreement of velocity magnitudes. Further quantitative comparison was carried out in this case study to show the goodness of the agreement between computational and experimental data (i.e. validation of the computational model).

4.3. Testing devices: the need for compliant 3D models

Experimental models can not only incorporate 3D-printed models but can also be used to test devices, e.g. a stent deployed into the 3D model itself. The system can, as per 4.2, be not only valuable for testing novel devices but also for generating validation data for computational model of the device itself. From a 3D printing perspective, in this case, it is interesting to consider what options are available in terms of 3D printing materials. While for other experiments or other applications and conventional materials such as SLA resins might be appropriate, in the case of deploying a device inside the 3D model, it may be desirable to implement the realistic compliance of blood vessels, allowing the device to adapt well to the anatomical implantation site. With advances in 3D printing technology, models such as silicone are becoming compatible with the printers, thus opening interesting avenues of research. One of the first 3D printing compatible compliant materials was a commercially available compound (TangoPlus FullCure) [27]. A study was carried out to evaluate the range of distensibility that can be implemented by models of varying thickness printed with such materials. The study showed that the material is suitable for implementing the distensibility of different arteries, in a range of thicknesses 0.7–1.5 mm. However, limitations of the materials include its fragility (i.e. tearing under pressure) and cost (more expensive that printing the same model in a rigid resin or nylon). The models for the compliance tests were 3D-printed segments of the descending aorta of a healthy volunteer, approximating a cylinder for ease of testing. Other anatomies were also 3D printed with the compliant compound. For instance, a study was carried out to test the implantation of a PPVI stent (see Section 3.1) in a patient-specific implantation site. By testing the device in a compliant model, it was possible to appreciate the way in which the stent adapted to the shape of the implantation site and deformed according to the anatomy (**Figure 7**). As per the validation framework (see Section 4.2), the same setting was also reproduced in silico [28].

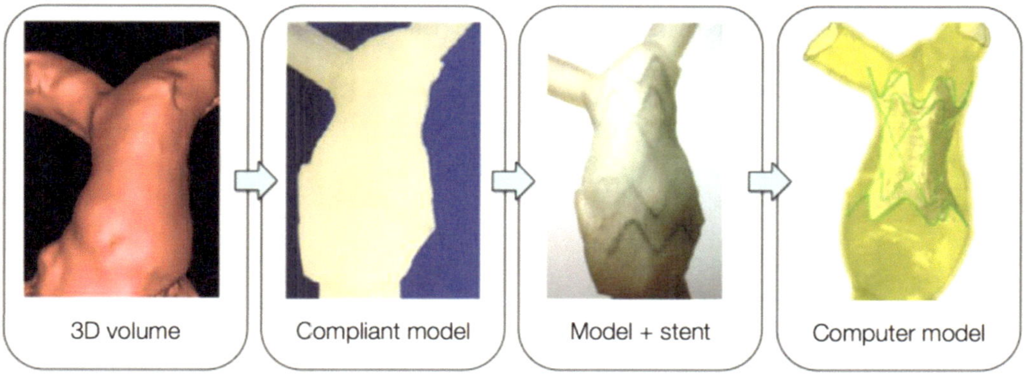

Figure 7. Reconstruction of patient-specific right ventricular outflow tract from CT data ('3D volume') which is 3D printed using a compliant rubber-like material ('compliant model'); a stent is inserted inside the 3D model for testing ('model + stent') and additional data can be acquired by replicating the 'model + stent' system computationally. Stent diameter at extremities = 40.7 mm.

Nowadays, TangoPlus FullCure is not the only option for 3D printing compliant materials for simulating arterial models. For example, another commercially available compound, namely Heart Print Flex [29], represents a valid material for 3D printing compliant cardiovascular models. Considerations of 3D printing compliant models include the wall thickness, which may represent a limit in terms of the printer resolution and in turn could restrict the range of distensibilities that can be implemented. Robustness of the models would also be desirable if these were to be used for hydrodynamic applications, e.g. testing devices in deformable models under pulsatile flow.

5. Facilitating doctor-patient communication

Communication between cardiologists/surgeons and patients with CHD and/or families of children with CHD is a delicate issue and an example of communication between an expert and a non-expert subject in which the information being exchanged is both sensitive and complex [30]. Studies have shown that understanding CHD and overall knowledge of the defect can be an issue in both patients, particularly adolescents, and their families [31]. Important information, such as the name of the primary diagnosis, has been shown to be difficult to retain in some instances, which could represent an issue for reporting information to other medical professionals caring for the patient and the patient closer community. Also, knowledge of the defect and overall appreciation of the anatomy can have an impact on lifestyle adjustments, for both the patient and the carer. In this light, improvements in communication with novel tools can be highly desirable. 3D models printed with rapid prototyping technology have been advocated to be potential communication tools; however, evidence in this context is lacking. As part of the translational work at the Centre for Cardio-vascular Imaging, models were purposefully manufactured for communicating with patients and families. A questionnaire-based study [32], in particular, was designed to assess parental knowledge of CHD, randomising parents into two groups: a) with 3D model and b) routine consultation without model. Models of a range of CHD were manufactured in white nylon, as a neutral choice to approach the parents. Examples of 3D models from this study are shown in **Figure 8**. The study provided some interesting findings. It was noted that using 3D models during clinical consultations in the setting of cardiac transition clinic lengthened consultation time by 5 minutes on average, possibly indicating a more thorough explanation. Clinicians rated the models as 'very useful' and reported that generally parents interacted well with the models. Parents themselves greatly appreciated the models and commented positively on their features. However, short-term knowledge, quantified by open-end answers but also through the use of diagrams and keywords (**Figure 9**), did not appear to improve across the two groups. When two blinded cardiologists were shown the parental responses to the questionnaire, they were unable to identify the primary diagnosis of the child in still approximately 40% of the cases. On the contrary, feedback provided insight into some interesting features of the models from the parents' perspective. Parents reported that medical images (e.g. echocardiography images, often times shown during consultation) are 'meaningless', while a 3D model 'makes it all so much easier for the non-expert to understand'. Parents reported some degree of shock in realising the patient-specific nature of the model ('[…] seeing the model of my son's heart was quite a shock') but understood and appreciated the value of the model for communicating anatomical features of the CHD ('[…] once I was used to the idea that the model I was looking at was what my son's heart actually looked like, it was a very useful tool for the doctor to illustrate exactly what was wrong with his heart').

Figure 8. Samples of models (not to scale) that are 3D printed in white nylon for assessing parental knowledge of CHD in a questionnaire-based study. All models are reconstructed from CMR data.

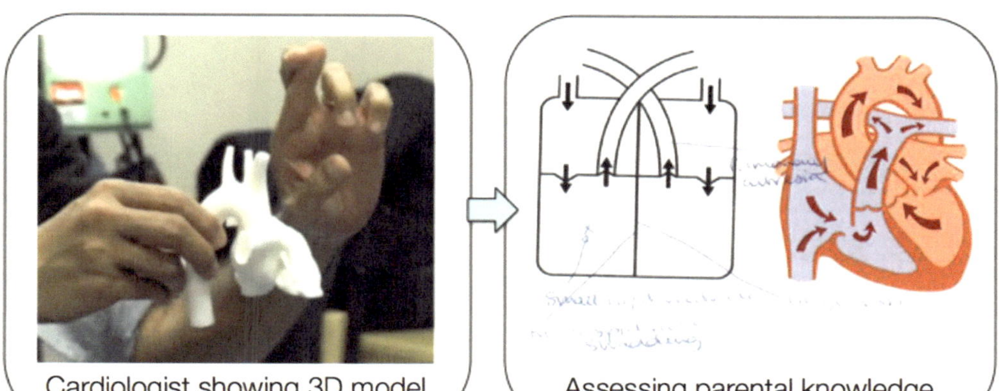

Figure 9. Cardiologists used 3D-printed models in discussing the anatomy and the CHD with the parents and the patients during their consultation. Parental knowledge was assessed before and after the consultation to assess possible improvements in short-term knowledge after having seen and manipulated the 3D model.

This line of work touches on important principles of patients & public involvement and engagement (PPI/E) in research. PPI, according to the INVOLVE national advisory group (part of the National Institute of Health Research), can be referred to as research that is carried out with/by the public, not about/for them, e.g. prioritising research topics, taking part in steering groups, undertaking research or disseminating research findings [33]. The importance of developing PPI is linked to the fact that research stemming from a PPI framework is likely to be more ethical and more relevant for the patients. The area of 3D printing cardiovascular models can be conducive to developing PPI, by actively involving patients' representatives in the process of improving not only model features or prioritising their use but also the communication around CHD and in turn promote a PPI framework through a true conversation with patients and their families (**Figure 10**).

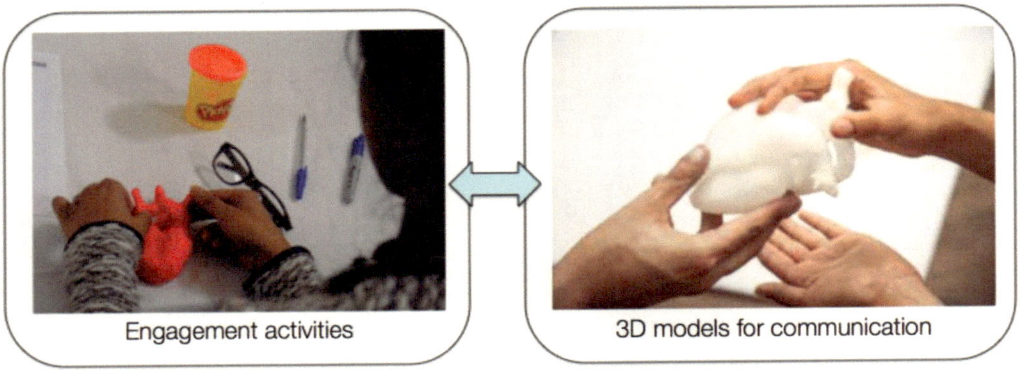

Figure 10. Engagement activities with patients and families can be carried out by a multidisciplinary team exploring concepts such as making a 3D heart (left: patient with CHD making a 3D heart with Play Doh®). Developing a conversation with the public can in turn inform better usage of 3D models and technology for communication purposes. Right image courtesy of Stephen King.

6. A multidisciplinary approach: incorporating 3D models in an art installation in a clinical setting

As part of the research activities of the Centre and of Great Ormond Street Hospital for Children, interactions with artist in residence Sofie Layton have resulted in an innovative and exciting use and display of 3D-printed models in the hospital/research setting. A multidisciplinary collaboration was established with the artist, developing an arts-and-science approach for the study of some CHD features. In particular, the artist was developing 'Under the Microscope', a project supported by the Wellcome Trust to explore how patients and families process medical information and medical language. While also exploring rare disease, Sofie Layton immersed herself in the world of CHD by familiarising with CMR images, features of the anatomy of different defects, running workshops with CHD patients and their parents, developing a conversation with cardiologists and nurses, and also appreciating engineering methods that can play a role in this context, such as computational simulations and 3D-printed models. From an artist's perspective, 3D printing is a fascinating concept and the exploration of forms that the technology allowed specifically for CHD was the starting point of the collaboration. In the culmination of the project, Sofie Layton realised two installations (exhibited at the Institute of Child Health and Great Ormond Street Hospital for Children,

http://dx.doi.org/10.5772/63411

February-April 2016). One of the installations revolves around CHD, including participatory work, a soundscape recounting the journey of a cardiac patient, panels that illustrate the language of CHD (i.e. anatomy, conditions, medications, treatments, narratives), as well as 3D-printed anatomical models (**Figure 11**).

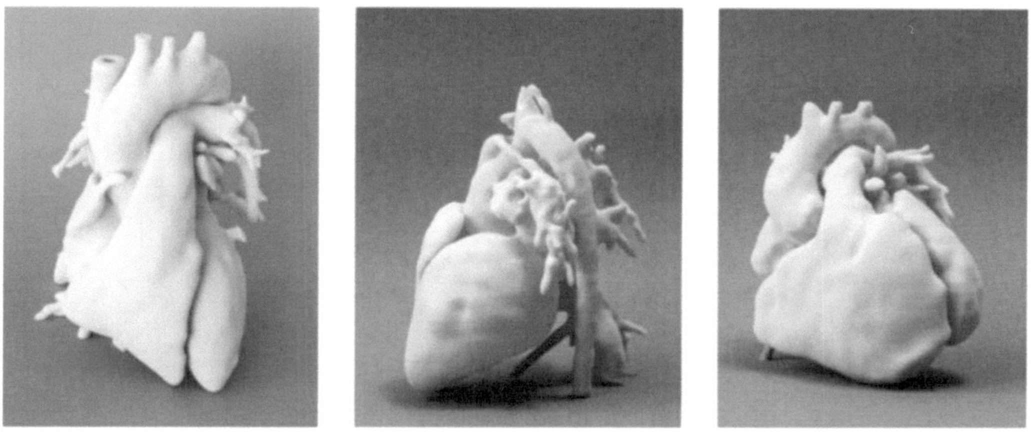

Figure 11. Examples of 3D models realised for Sofie Layton's 'Under the Microscope' art installation, including a normal heart (left), a case of repaired TGA with ASO (centre), a case of repaired tetralogy of Fallot (right). Image courtesy of Stephen King.

In this context, models lend themselves to contribute to this site-specific installation by representing the variety of forms and arrangements that the cardiovascular anatomy can present, but the artist can take the process one step further. By exploring shapes and sizes, some unconventional models are included in the landscape (see **Figure 12**). A small bronze heart, suspended in a bell jar, is the result of scaling the 3D volume of a normal heart model, 3D printing a mould and casting a precious heart, sparking a reflection on the preciousness of the organ and of the stories and symbols associated with it. Similarly, another model, also 3D prototyped, has been virtually manipulated prior to printing so to resemble a Rubick's cube, which emerged from a conversation with a patient as a powerful image of the complexity of his heart following multiple repairs of CHD. The models are positioned on a sterile-looking stainless steel table in amongst the other artwork, essentially creating a dialogue between medical conditions and real-life stories. This collaboration is an example of how 3D printing technology allows for a stimulating exploration of possibilities of expression, which is facilitated and promoted by the continuous dialogue between the artist and the engineer, in a sense pushing the boundary of the technology for its medical use. The piece, indeed, has a clinical relevance, as well as relevance for the abovementioned concept of PPI/E. By creating this landscape (including sounds) in a medical/research setting rather than a neutral space, the installation echoes conversations and stories that have taken place within that dimension but enriches them by allowing for one final element to emerge: the human element. It is the artist who takes the responsibility of blending real-life stories in between anatomical models, technical jargon and heart outlines from CMR. This innovative approach presents yet one more use for 3D-printed models and for the possibility of expression offered by this technology.

Figure 12. Installation of 3D models under bell jars, as part of Sofie Layton's piece 'Under the Microscope'. While most models have a medical appearance, 3D technology is allowed to manipulate the data and scale it to create, for instance, a small bronze heart (right) which the artist chose to suspend in one of the bell jars. Image courtesy of Stephen King.

7. Outlook and considerations for future research

Based on a single-centre experience, 3D printing can present several relevant uses for the study and communication around CHD. Such a revolutionary manufacturing technique allows to generate, in a relatively short time and at a relatively affordable cost, patient-specific models, which—as discussed— are particularly relevant when referring to unique anatomical arrangements, as often the case in CHD. Based on our Centre's experience, as briefly outlined in this chapter, uses range from testing novel devices and appreciating patient suitability for a specific device, providing additional insight into a particular anatomy, allowing for performing parametric studies and validating computational models, as well as representing a medium for implementing PPI/E in this area of medicine. A few considerations are necessary with regard to the future uses of 3D printing for CHD, including

- Image availability: the creation of 3D models is, of course, dependent on the availability of imaging data (CMR, CT) and of 3D datasets within the imaging protocol; this should be taken into account when considering which patients could potentially receive a 3D model or for which cases a 3D model could be created, e.g. for aiding decision-making (i.e. if there is no clinical indication for imaging, then a model will not be available for that case).
- Operators: image segmentation and reconstruction does currently still require a specialised operator, and it can be a time-consuming task depending on the complexity of the defect(s) and imaging quality; while ours is a specialist Centre in CHD and has a team on engineers in situ, other centres may not have the capabilities of handling the data and/or manufacturing the models, which can be a hurdle towards clinical translation;
- Cost: depending on the volume of the part and the material, models can cost in the range of £30–£500. When considering a large volume of models, for potentially translating the technique to the point of even becoming a service, an appropriate cost-effectiveness analysis should be carried out on a large scale, including not only the printing costs but also the man time for image elaboration, which is integral to the process. This element is still lacking.
- Printing time: a full heart model of an adolescent would typically require around 6–7 hours to print (or left to print overnight); however, recent advances on the technology side have led to 3D printers that boast drastically quicker printing times than current technologies. The possibilities of fast printing could be appealing for clinical translation, depending on the cost.
- Efficacy: regardless of the cost, clinical translation of 3D printing models in medicine and in CHD in particular still depends on the demonstration of its efficacy. Considering the rarity of some CHDs, this requires a multi-centre effort. While initiatives such as the NIH 3D Print Exchange (Heart Library, http://3dprint.nih.gov/collections/heart-library) bring together a large number of models, a systematic quantification with statistical evidence of the usefulness of the models in clinical practice is also lacking at present.
- Other applications: other uses of 3D models are currently being explored beyond those mentioned in this chapter, particularly including the use of 3D models of CHD for training purposes (e.g. nurses or students); the use of 3D-printed models for training has been used in other areas of medicine [34] and also, in CHD, in a study that focused on the appreciation of ventricular septal defects in a cohort of medical students [35] suggesting the beneficial use of 3D models for better appreciation of the defect.
- Availability of the technology in situ: it is very rare that a clinical centre would have 3D printing facilities, at the moment; research hospitals would likely have access to the facilities via academic collaborators. This might change in the near future, depending on the development on new machines, possibly including high-definition desktop 3D printers, which are affordable and do not require a specialised dedicated room, depending on the purpose for which the model is required.
- Collaboration: as for any translational effort, collaboration is of paramount importance between different users and between different specialists, in this case particularly between

engineers (on the technical side) and clinicians (on the user side). The patients' perspective is also important, especially for using models for communication purposes, and a PPI/E framework should be advocated for such applications.

Acknowledgements

The authors acknowledge the generous support of the Royal Academy of Engineering, National Institute for Health Research, and Heart Research UK. The authors are also grateful to artist Sofie Layton and photographer Stephen King for their collaboration.

Author details

Giovanni Biglino[1,2*], Claudio Capelli[1], Andrew M. Taylor[1] and Silvia Schievano[1]

*Address all correspondence to: g.biglino@bristol.ac.uk

1 Centre for Cardiovascular Imaging, Institute of Cardiovascular Science, University College London & Great Ormond Street Hospital or Children, NHS Foundation Trust, London, United Kingdom

2 Bristol Heart Institute, School of Clinical Sciences, University of Bristol, Bristol, United Kingdom

References

[1] Lipson H. 3D printing: the technology that changes everything. New Scientist. 2011; 2823.

[2] D 'Aveni R. The 3-D printing revolution. Harvard Business Review. Harvard Business Publishing, Boston, MA, USA 2015.

[3] Munoz C, Kim C, Armstrong L. Layer-by-layer: opportunities in 3D printing technology trends, growth drivers and the emergence of innovative applications in 3D printing. MaRS Market Insights. [Internet]. 2013. Available from: https://www.marsdd.com/wp-content/uploads/2014/04/MAR-CLT6965_3D-Printing_White_paper.pdf [Accessed 2016-03-07]

[4] Alford J. 3d printed implants save the lives of 3 babies. IFL Science. [Internet] 2015. Available from: http://www.iflscience.com/health-and-medicine/how-3d-printing-transformed-lives-3-babies-life-threatening-diseases [Accessed 2016-03-07]

[5] Owusu JA, Boahene K. Update of patient-specific maxillofacial implant. Curr Opin Otolaryngol Head Neck Surg. 2015;23:261–4. DOI: 10.1097/MOO.0000000000000175

[6] Chung KJ, Hong do Y, Kim YT, Yang I, Park YW, Kim HN. Preshaping plates for minimally invasive fixation of calcaneal fractures using a real-size 3D-printed model as a preoperative and intraoperative tool. Foot Ankle Int. 2014;35:1231–6. DOI: 10.1177/1071100714544522

[7] Darwood A, Collier J, Joshi N, Grant WE, Sauret-Jackson V, Richards R, Dawood A, Kirkpatrick N. Re-thinking 3D printing: A novel approach to guided facial contouring. J Craniomaxillofac Surg. 2015;43:1256–60. DOI: 10.1016/j.jcms.2015.06.001

[8] Schievano S, Sebire NJ, Robertson NJ, Taylor AM, Thayyil S. Reconstruction of fetal and infant anatomy using rapid prototyping of post-mortem MR images. Insights Imaging. 2010;1:281–286, ISSN 1869-4101.

[9] Powers MK, Lee BR, Silberstein J. Three-dimensional printing of surgical anatomy. Curr Opin Urol. 2016. [Epub ahead of print]Curr Opin Urol 2016;26(3):283-8. doi: 10.1097/MOU.0000000000000274.

[10] Shirakawa T, Koyama Y, Mizoguchi H, Yoshitatsu M. Morphological analysis and preoperative simulation of a double-chambered right ventricle using 3-dimensional printing technology. Interact Cardiovasc Thorac Surg. 2016. [Epub ahead of print] Interact Cardiovasc Thorac Surg 2016;22(5):688-90. doi: 10.1093/icvts/ivw009

[11] Biglino G, Schievano S, Taylor AM. The use of rapid prototyping in clinical applications. In: Hoque ME, editor. Applications of rapid prototyping technology in modern engineering. Rijeka: InTech; 2011. DOI: 10.5772/24128

[12] Coats L, Khambadkone S, Derrick G, Sridharan S, Schievano S, Mist B, Jones R, Deanfield JE, Pellerin D, Bonhoeffer P, Taylor AM. Physiological and clinical consequences of relief of right ventricular outflow tract obstruction late after repair of congenital heart defects. Circulation. 2006;113:2037–44.

[13] Schievano S, Petrini L, Migliavacca F, Coats L, Nordmeyer J, Lurz P, Khambadkone S, Taylor AM, Dubini G, Bonhoeffer P. Finite element analysis of stent deployment: understanding stent fracture in percutaneous pulmonary valve implantation. J Interv Cardiol. 2007;20:546–54.

[14] Schievano S, Migliavacca F, Coats L, Khambadkone S, Carminati M, Wilson N, Deanfield JE, Bonhoeffer P, Taylor AM. Percutaneous pulmonary valve implantation based on rapid prototyping of right ventricular outflow tract and pulmonary trunk from MR data. Radiology. 2007;242:490–7.

[15] Schievano S, Taylor AM, Capelli C, Coats L, Walker F, Lurz P, Nordmeyer J, Khambadkone S, Tsang V, Carminati M, Bonhoeffer P. First-in-man implantation of a novel percutaneous valve: a new approach to medical device development. EuroIntervention. 2010;5:745–50.

[16] Cosentino D, Capelli C, Derrick G, Khambadkone S, Muthurangu V, Taylor AM, Schievano S. Patient-specific computational models to support interventional procedures: a case study of complex aortic re-coarctation. EuroIntervention. 2015;11:669–72. DOI: 10.4244/EIJY15M09_03.

[17] Butts RJ, Hsia TY, Hamilton Baker G. Feasibility of conductance catheter-derived pressure-volume loops to investigate ventricular mechanics in shunted single ventricles. Cardiol Young. 2013;23:776–9. DOI: 10.1017/S1047951112002053

[18] Biglino G, Giardini A, Baker C, Figliola RS, Hsia TY, Taylor AM, Schievano S. In vitro study of the Norwood palliation: a patient-specific mock circulatory system. ASAIO J. 2012;58:25–31. DOI: 10.1097/MAT.0b013e3182396847

[19] Biglino G, Giardini A, Baker C, Figliola RS, Hsia TY, Taylor AM, Schievano S. Implementing the Sano modification in an experimental model of first-stage palliation of hypoplastic left heart syndrome. ASAIO J. 2013;59:86–9. DOI: 10.1097/MAT.0b013e3182768b7f

[20] Quarteroni A, Veneziani A. Analysis of a geometrical multiscale model based on the coupling of PDE's and ODE's for blood flow simulations. Multiscale Model Simulat: a SIAM Interdisciplinary Journal. 2003;1:173–95. ISSN 1540-3459

[21] Figliola RS, Giardini A, Conover T, Camp TA, Biglino G, Chiulli J, Hsia TY. In vitro simulation and validation of the circulation with congenital heart defects. Prog Pediatr Cardiol. 2010;30:71–80.

[22] Feinstein JA1, Benson DW, Dubin AM, Cohen MS, Maxey DM, Mahle WT, Pahl E, Villafañe J, Bhatt AB, Peng LF, Johnson BA, Marsden AL, Daniels CJ, Rudd NA, Caldarone CA, Mussatto KA, Morales DL, Ivy DD, Gaynor JW, Tweddell JS, Deal BJ, Furck AK, Rosenthal GL, Ohye RG, Ghanayem NS, Cheatham JP, Tworetzky W, Martin GR. Hypoplastic left heart syndrome: current considerations and expectations. J Am Coll Cardiol. 2012;59:S1–42. DOI: 10.1016/j.jacc.2011.09.022

[23] Vukicevic M, Chiulli JA, Conover T, Pennati G, Hsia TY, Figliola RS. Mock circulatory system of the Fontan circulation to study respiration effects on venous flow behavior. ASAIO J. 2013;59:253–60. DOI: 10.1097/MAT.0b013e318288a2ab

[24] Taylor CA, Figueroa CA. Patient-specific modeling of cardiovascular mechanics. Annu Rev Biomed Eng. 2009;11:109–34. DOI: 10.1146/annurev.bioeng.10.061807.160521

[25] Kung EO, Les AS, Figueroa CA, Medina F, Arcaute K, Wicker RB, McConnell MV, Taylor CA. In vitro validation of finite element analysis of blood flow in deformable models. Ann Biomed Eng. 2011;39:1947–60. DOI: 10.1007/s10439-011-0284-7

[26] Biglino G, Cosentino D, Steeden JA, De Nova L, Castelli M, Ntsinjana H, Pennati G, Taylor AM, Schievano S. Using 4D cardiovascular magnetic resonance imaging to validate computational fluid dynamics: A case study. Front Pediatr. 2015;3:107. DOI: 10.3389/fped.2015.00107

[27] Biglino G, Verschueren P, Zegels R, Taylor AM, Schievano S. Rapid prototyping compliant arterial phantoms for in-vitro studies and device testing. J Cardiovasc Magn Reson. 2013;15:2. DOI: 10.1186/1532-429X-15-2

[28] Biglino G, Capelli C, Binazzi A, Reggiani R, Cosentino D, Migliavacca F, Bonhoeffer P, Taylor AM, Schievano S. Virtual and real bench testing of a new percutaneous valve device: a case study. EuroIntervention. 2012;8:120–8. DOI: 10.4244/EIJV8I1A19

[29] Material characterization of Materialise HeartPrint® models and comparison with arterial tissue properties [Internet] 2016. Available from: http://biomedical.materialise.com/white-papers-material-characteristics-heartprint-flex-models-abstract [Accessed 2016-03-07].

[30] Bigi S. The persuasive role of ethos in doctor-patient interactions. Commun Med. 2011;8:67–76. DOI:10.1558/cam.v8i1.67

[31] Cheuk DK, Wong SM, Choi YP, Chau AK, Cheung YF. Parents' understanding of their child's congenital heart disease. Heart. 2004;90:435–9. DOI:10.1136/hrt.2003.014092

[32] Biglino G, Capelli C, Wray J, Schievano S, Leaver LK, Khambadkone S, Giardini A, Derrick G, Jones A, Taylor AM. 3D-manufactured patient-specific models of congenital heart defects for communication in clinical practice: feasibility and acceptability. BMJ Open. 2015;5:e007165. DOI: 10.1136/bmjopen-2014-007165

[33] Involve. Briefing note two: What is public involvement in research? [Internet]. 2015. Available from: http://www.invo.org.uk/posttyperesource/what-is-public-involvement-in-research/ [Accessed 2016-03-07]

[34] Adams JW, Paxton L, Dawes K, Burlak K, Quayle M, McMenamin PG. 3D printed reproductions of orbital dissections: a novel mode of visualising anatomy for trainees in ophthalmology or optometry. Br J Ophthalmol. 2015;99:1162–7. DOI: 10.1136/bjophthalmol-2014-306189

[35] Costello JP, Olivieri LJ, Krieger A, Thabit O, Marshall MB, Yoo SJ, Kim PC, Jonas RA, Nath DS. Utilizing three-dimensional printing technology to assess the feasibility of high-fidelity synthetic ventricular septal defect models for simulation in medical education. World J Pediatr Congenit Heart Surg. 2014;5:421–426.

Chapter 7

Regenerative Repair of Bone Defects with Osteoinductive Hydroxyapatite Fabricated to Match the Defect and Implanted with CAD, CAM, and Computer-Assisted Surgery Systems

Koichi Yano, Takashi Namikawa, Takuya Uemura,
Yasunori Kaneshiro and Kunio Takaoka

Additional information is available at the end of the chapter

http://dx.doi.org/10.5772/63743

Abstract

Regenerative repair of large bone defects currently remains a challenging issue during surgery, owing to the limited regenerative ability of the bone. To address this issue, we attempted a precise repair of a bone defect using computer-aided procedures. Using pelvic computed tomography (CT) images of beagle dogs, virtual tumors were created in the pelvis using computer-aided design (CAD), and a bone resection following the margins of the bone tumor was performed on the CAD image. Hydroxyapatite (HA) implants to fill the bone defects and implants for shape evaluation of bone resection sites were designed and produced by computer-aided manufacturing and three-dimensional printing. Subsequently, using a computer navigation system, iliac bone defects were created in beagle dogs as preoperatively planned on CAD, filled with HA implants shaped to fit the bone defect sites, and coated with a recombinant human bone morphogenetic protein (rhBMP)-2-containing dough bone-forming material. Postoperative CT revealed that the new bone was formed around the implant over time. Anatomical healthy bone repair was confirmed to be completed 12 weeks after the surgery. These results demonstrate potential novel technology for efficacious and accurate repair of large bone defects without bone grafting.

Keywords: regenerative repair, bone defect, computer-aided design, computer-assisted surgery, 3D printing

1. Introduction

Extensive bone defects can occur after severe trauma, infection, or bone tumor resection, and in some cases require bone tissue reconstruction. Therefore, auto/allografts and artificial materials are implanted. Materials currently used for bone tissue reconstruction include autologous bone tissue, such as the ilium and fibula, and allogeneic materials, such as cryopreserved bone, titanium alloys, and bioactive ceramics [1]. Each of these has distinct advantages, but they also have various disadvantages or problems that remain to be solved.

The autologous bone is the most effective material for small bone defects and is characterized by strong bone-forming ability, accompanied by the capability of bone union and remodeling capacity. However, it has disadvantages, including limitations in the collectable quantity and complications after collection, such as pain, infection, fracture, deformation, and risk of damage to major nerves or blood vessels. The use of allogeneic bone is associated with low bone-forming ability; potential transmission of infectious agents; and cost problems, including cleanliness management. Disadvantages of the heterologous bone include possible transmission of animal infections and immunological rejection in addition to those listed for allogeneic bone [2–4].

Although artificial materials offer the advantages of easy access and processing, they are usually incapable of practical bone formation and thus are ineffective for bone regeneration in large bone defects. It is difficult to reconstruct relatively large bone defects in a shape that is anatomically similar to the normal structure.

Scaffolds, growth factors, and cells represent three key elements in regenerative medicine. An ideal approach for bone repair with regenerative medicine technology should have the abilities of osteogenesis, osteoconduction, osteoinduction, and osteointegration to resolve the disadvantages of currently available graft materials. Biocompatible and biodegradable scaffolds include those made of biological materials, such as type 1 collagen and demineralized bone, and of synthetic (artificial) materials, such as porous metals, bioactive glass, synthetic polymers, and calcium phosphate ceramics [hydroxyapatite (HA) and tricalcium phosphate (TCP)] [5].

Among bone morphogenetic proteins (BMP), recombinant human BMP (rhBMP)-2 and -7 are potent osteoinductive cytokines and are clinically used as graft materials in spinal fusion, pseudarthrosis after long bone fractures, repeated posterolateral fusion, and treatment of open fractures of the tibia in some countries, such as European countries and the United States. We expect that they will find increasing application in bone-regenerating medicine because of their high rates of bone union and their simplicity and minimal invasiveness (i.e., bone harvesting is not required) [6–8].

Recent advances in computer-assisted techniques have led to computer-assisted preoperative planning, custom production of surgical implants using patient data, and use of navigation systems in orthopedics. Examples include three-dimensional (3D) printing-based preoperative planning, in which 3D printing of the relevant bone is performed before a trauma patient undergoes surgery and plates to be used are templated with the reference to the simulated bone so that no intraoperative plate bending maneuvers are required. Furthermore, the pre-bent plate itself serves as a reduction indicator, and accurate corrective osteotomy procedures for malunited fractures are prepared for each patient using osteotomy guides [9, 10]. In addition, the use of navigation systems has been reported to be useful in joint surgery, for example, for accurate installation of artificial joint components, safe and accurate insertion of pedicle screws in spinal fusion, and resection of musculoskeletal tumors [11–13].

These computer-assisted imaging technologies might be applicable for repair of large bone defects that result from wide resection of malignant musculoskeletal tumors. In those cases, the decision regarding optimal resection margins should be based on 3D computed tomography (CT) or/and magnetic resonance imaging (MRI) data to avoid exposure of the tumor tissue to the surgical field and late local recurrence of the tumor, and the image data would be applicable to surgical procedures using computer-assisted navigation surgery (CAS) system. Image data of the virtual bone defect in the computer-aided design (CAD) would be used to fabricate a porous HA block implant with computer-assisted manufacturing system (CAM) to fill the bone defect. To add osteoinductive capacity to the HA implant to accelerate bone defect repair, a cytokine rhBMP-2 with potent osteoinductive capacity retained in its dough-like delivery system is pasted on the HA surface during implantation.

In this study, we investigated the feasibility of computer-assisted technology (CAD, CAM, and CAS) for preoperative surgical planning and corrective surgical procedures to resect a virtual tumor and fabricate an HA implant. Another aim was to estimate the efficacy of rhBMP-2 and its delivery system to promote repair of a large bone defect without bone grafting in a beagle model.

2. Experiments

2.1. Methods

2.1.1. Preoperative planning

In beagle dogs (male, 10-month-old, body weight: 9–11 kg) scheduled for surgery, pelvic CTs were performed a few days before the surgery, and the acquired data [digital imaging and communication in medicine (DICOM)] were used for preoperative planning with CAD software (Mimics, Magics, Materialise, NV, USA). 3D images created from CT data after DICOM data are transferred to CAD and converted into a standard triangulated language format that can be separated or combined in any way with CAD. On the 3D images of the pelvis obtained in this manner, a virtual bone tumor of 15 mm in diameter was created in the left iliac wing, and a bone resection model was created by establishing the bone resection line, 10 mm distant from the tumor (**Figure 1a** and **b**). An implant was designed to compensate for the defect occurring after resection by subtracting the bone defect part from a mirror image of the right ilium (**Figure 2a**).

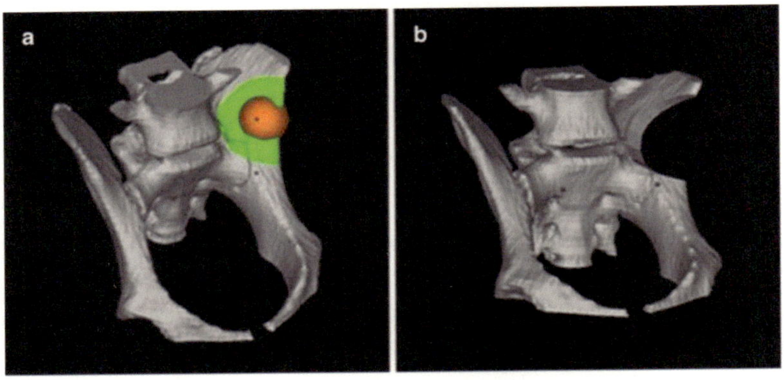

Figure 1. (a) A virtual spherical bone tumor of 15 mm in diameter (color: orange) was created in the left iliac wing of the canine pelvis with preoperative planning on computer-aided design (CAD) software. The bone resection line was set 10 mm distant from the bone tumor on the CAD image (color: green). (b) A bone resection model was established on CAD image.

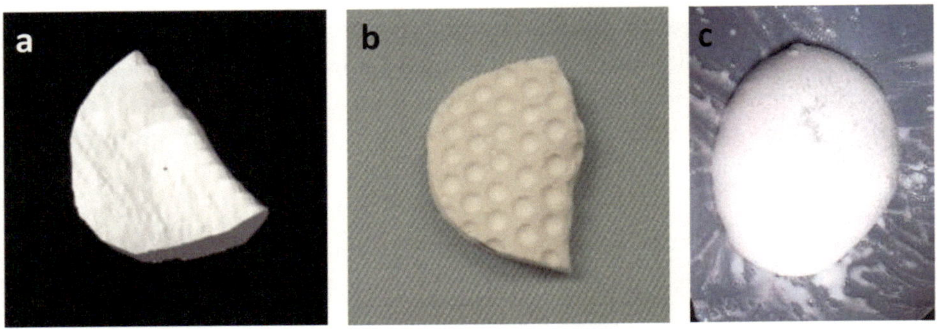

Figure 2. (a) CAD image of defected bone generated by subtracting the left iliac bone from the mirror image of the right iliac bone. (b) Fabricated interconnected porous calcium hydroxyapatite (IP-CHA) block by using a 3D drilling machine based on CAD data. Multiple dimples were made to facilitate retention of the BMP-retaining dough material. (c) BMP-retaining biodegradable dough material.

2.1.2. Implant preparation

We selected an artificial bone made of interconnected porous calcium HA (IP-CHA, pore size: 150 μm, porosity: 72–78%, MMT Co., Ltd., Osaka, Japan) as a material to fill the defect because of its high osteoconductive capacity [14]. CT data of the bone defect as described above were transferred to a 3D cutting machine (CAM; Modela player, Roland DG Co., Shizuoka, Japan), and an IP-CHA block (cuboid, 40 × 20 × 10 mm) was processed with a 3D drilling machine (MDZ-20, Roland DG Co.). The processed HA block was then machined with a high-speed surgical drill to create a large number of dimples facilitating the application of a BMP-containing dough bone-forming material (described below) (**Figure 2b**). To confirm the size of the defect at the bone resection site during surgery, CAD design data were migrated to a 3D printer (ZPrinter Z-310, Toyotsu machinery, Aichi, Japan) and printed with gypsum. These were autoclaved or gas sterilized so that they could be used during surgery.

2.1.3. Dough bone-forming material

Because BMP administered alone rapidly diffuses in vivo, it cannot maintain the local concentration at a level required for bone formation and thus does not induce bone formation. Therefore, an appropriate slow-release carrier (drug delivery system, DDS) is necessary to accomplish effective bone formation with BMP. Bovine type 1 collagen is currently used as a DDS for BMP, but it has the disadvantages of requiring a large amount of expensive BMP and the potential transmission of pathogens, such as variant Creutzfeldt-Jakob disease, which cannot be completely excluded [4].

To solve these problems, our group has developed a poly-d, l-lactic acid-polyethylene glycol block copolymer (PLA-PEG, Taki Chemical Co., Ltd., Kakogawa, Japan), which is a fully synthetic artificial DDS material that reduces the BMP dose required for bone formation [15–25]. Given that the PLA-PEG polymer is viscous at room temperature, it is difficult to process manually but can be transformed into a dough-like material by the addition of powdery β-TCP, which can be readily processed. By adding BMP to this mixture, we devised a dough-like bone-forming material (**Figure 2c**) [26–32]. rhBMP-2 was synthesized by *Escherichia coli* transfected with the rhBMP-2 gene. After inactive monomer rhBMP-2 molecules were produced, they were dimerized to active rhBMP-2 by biochemical processing. The BMP-2 delivery system for one IP-CHA implant was generated by mixing 100 μg of rhBMP-2 (Osteopharma Inc., Osaka, Japan), 200 mg of powdery β-TCP (particle size <100 μm in diameter, Olympus Biomaterial Corp., Tokyo, Japan), and 200 mg of the PLA-PEG polymer [33]. Physiochemical characteristics of this dough bone-forming material have been reported previously. Dough bone-forming material preparations devoid of rhBMP-2 were prepared as the control group. The samples were stored at −30°C until use during surgery.

2.2. Bone resection with the navigation system and implantation of BMP-added IP-CHA implant

Data of the bone resection model created with CAD software were converted into the DICOM format. Converted DICOM data were transferred to a CT-based computer navigation system (Stealth Station Tria, Medtronic Navigation, Louisville, Co, USA). Anesthesia, ketamine (10 mg/kg) and xylazine (1.2 mg/kg) was injected intramuscularly and maintained with a continuous pentobarbital (25 mg/kg) intravenously. The left ilium was opened at the acetabular margin in beagle dogs in the right lateral decubitus position. A reference frame was secured to the iliac wing with two threaded pins and matched with navigation by point and surface registration. A Surgairtome equipped with a passive pointer was used for bone resection. Bone resection was performed as preoperatively planned according to the navigation system. The accuracy of bone resection was evaluated by testing the fit of the 3D printer-produced gypsum implant. The dough bone-forming material was then pasted evenly on the surface of the preoperatively designed and processed HA implant, and the coated implant was secured with two or three 0.6 mm wires (Synthes Co., Ltd., Tokyo, Japan). The implant was carefully sutured in place to ensure contact with the surrounding muscle. A prophylactic antibiotic agent (10 mg/kg body weight) was administered to animals before and after the surgery. This animal experiment was performed after institutional approval was obtained.

2.3. Evaluation of efficacy of computer-assisted bone defect repairing system

CT slice images of the iliac bone were obtained immediately and 3, 6, 9, and 12 weeks postoperatively to observe the bone repair reaction at the bone defect site and reconstructed into slice and 3D images as desired using the reconstruction software (Aze, Tokyo, Japan). CT images of the mid-axial slice of the implant and the next slice 1 mm from the center were analyzed. A high-density area surrounding the implant was considered to be a new bone area and quantified with ImageJ software (Wayne Rasband, National Institutes of Health, Bethesda, MD). Results were statistically analyzed by the Mann-Whitney U test with a significance level of $P < 0.05$. The ilium was harvested 12 weeks postoperatively and fixed in 70% ethanol. From un-decalcified specimens, 10 μm sections were prepared and subjected to Villanueva bone staining for histological evaluation. Specimens were macroscopically evaluated.

3. Results

3.1. Accuracy of iliac bone resection by computer-aided surgery

In all animals, navigation-guided osteotomy could be accurately performed on the virtual bone tumor exactly as planned on the CAD system. The CAD-designed, 3D printer-produced

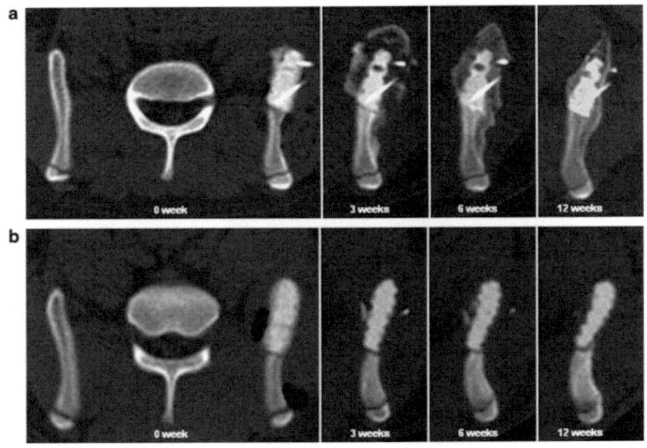

Figure 3. Representative mid-axial plane images of computed tomography (CT) sequentially taken over the experimental period. (a) In the BMP-treated group, radiopaque images were observed on the IP-CHA surface 3 weeks postoperatively. The radiologically dense area around the implant increased in a time-dependent manner. The implant was completely covered by new bone at 12 weeks postoperatively. (b) No significant radiopaque images were observed in the control group up to 12 weeks postoperatively.

implant fitted into the bone defect, and the IP-CHA implant prepared by CAD and CAM fitted well in all experimental and control animals. Over the 12-week study period, all animals survived without any complication.

3.2. Evaluation of bone defect repair using reconstructed CT images

Figure 3 shows representative mid-axial plane images sequentially taken over the experimental period in the BMP-treated group and no-BMP control group. In the BMP-treated group, in which the IP-CHA surface was coated with the BMP-retaining dough material, radiopaque images were observed on the implant surface 3 weeks postoperatively. Thereafter, the radiologically dense area around the implant increased in a time-dependent manner, and the implant was completely covered by radiopaque images at 12 weeks postoperatively. In contrast, no significant radiopaque images were observed in the control group images up to 12 weeks postoperatively.

3.3. Macroscopic and histological findings

Macroscopic findings for the BMP-treated group of animals indicated that the whole implant was completely encased by a new bone, and the new bone was continuous with the original ilium, suggesting successful repair of the bone defect to the normal anatomical morphology

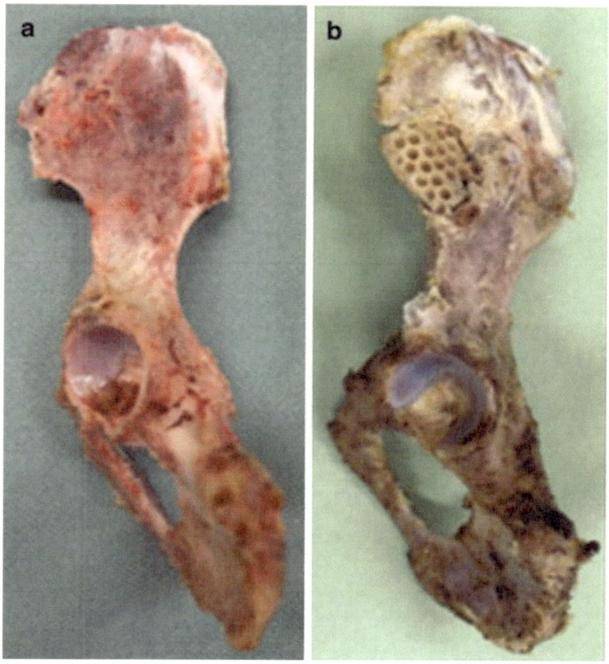

Figure 4. (a) BMP-treated group. The whole implant was completely encased by a new bone continuous with original ilium on macroscopic findings. (b) No bone tissue was observed on the implant surface in the control group.

(**Figure 4a**). In contrast, no bone tissue was observed on the implant surface in the control group of animals (**Figure 4b**). Representative histological specimens of un-decalcified tissues with Villanueva bone staining are shown (**Figure 5a–d**). In the BMP-treated group, the implant was covered by a new bone that was continuous with the original ilium, and the new bone was found to have entered pores inside the implant (**Figure 5c**). The dough material pasted on the implant surface was not observed in histological specimens obtained 12 weeks postoperatively. This indicates biodegradation during the experimental period. No inflammatory response was observed around the implant over the experimental period.

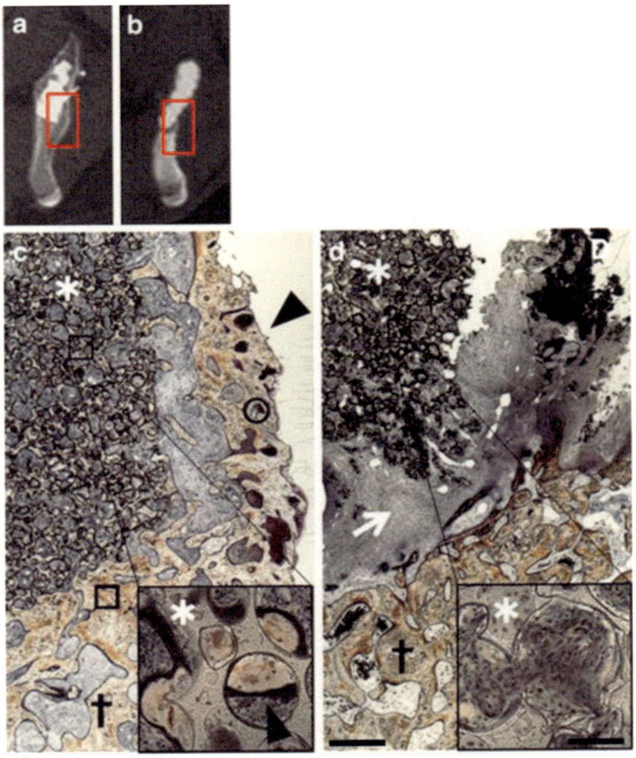

Figure 5. Histological sections at 12 weeks after surgery [lower (*20) and higher magnification (*200), *scale bar* indicates 100μm, 8μm in each magnification]. (a, b) Boxed area of reconstructed axial CT image shows a histological picture of (c) (BMP-treated group) and (d) (control group), respectively. (c) In the BMP-treated group, the implant was covered by a new bone that was continuous with the original ilium, and the new bone was observed in the pore of the implant. d. In the control group, no bone tissue was observed. (arrowhead: new bone, asterisk: IP-CHA implant, cross: ilium, arrow: fibrous tissue, circle: osteoid, box: calcified bone).

3.4. Changes over time in the amount of the new bone formed

In the BMP-treated group, the volume of the calcified new bone tissue around the implant on CT slice images of the pelvis peaked 3 weeks postoperatively, then decreased over time, and finally resulted in the anatomical shape of the original ilium in 12 weeks postoperatively (**Figure 6**). In contrast, significant new bone formation was not observed in the control group during the observation period.

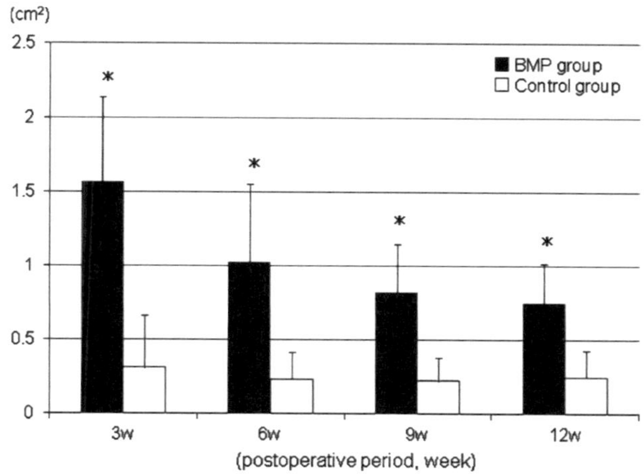

Figure 6. The amount of new bone formation of the BMP-treated group was significantly larger than that of the control group in reconstructed mid-axial CT planes at all experimental periods (statistical significant difference was set at $p<0.05$, values showed mean ± standard deviation).

4. Discussion

BMPs are established osteoinductive proteins, and BMP-2 and BMP-7, in particular, are used in combination with animal-derived collagen as a carrier for spinal fusion, pseudarthrosis treatment, and open fractures of the tibia [6–8]. These cytokine therapies accomplish a high rate of bone union and have reduced the frequency of bone grafting. However, the currently available carrier material of BMP is collagen derived from animals, which carries a possibility of infectious pathogens. We established a new DDS of BMP-2; efficacy of the rhBMP-2-retaining putty for repair of the large bone defect had been presented previously by our group in other bone defect models in beagles and posterolateral spinal fusion models in rabbits. In those experimental studies, a special A-B-A type of block copolymer (molecular weight 9,800, A;-*dl polyethylene glycol* and B; *polylactic acid)* was successfully used as the carrier material for BMP-2. The polymer was originally produced by us in collaboration with a chemical company as an optimal delivery system for rhBMP-2. Because this polymer itself was highly sticky and difficult to handle, an equal weight of biodegradable β-TCP powder was mixed with the polymer to transform it to a slightly sticky putty or dough for easy implantation. Thus far, with respect to the use of this carrier material in experimental animals, no local or systemic adverse events were noted, and no significant inflammatory reaction has been observed on histology. This synthetic carrier material is a candidate for efficacious use of rhBMP-2 to repair or reconstruct the damaged bone after extensive safety check and clinical trials. Regarding the rhBMP-2 dose added to the implant, 100 μg of rhBMP-2 in combination with 400 mg of the delivery system (carrier material) per animal was used in the current study. This formulation was fixed based on our previous experimental results obtained in beagle models. As widely known, a minimal dose of BMP-2 to elicit local new bone formation in *in vivo* condition depends on the animal species, and humans are less sensitive to BMP-2. Therefore, a higher BMP-2 dose to the order of a few mg will be required for bone defect repair as observed in clinical reports of rhBMP-2 used for spinal body fusion, non-union of fracture, and other off-label use of rhBMP-2 in humans.

Recent reports have described successful reconstruction of maxillary bone defects in rabbits and bone union achievement in rats with pseudarthrosis using a combination of BMP gene-transfected cells and an artificial material or using sheets of cultured mesenchymal stem cells [34, 35]. However, an expensive facility equipped with a special culture system is required to establish a cell culture system for clinical application in humans. Conversely, BMPs can be conveniently used anywhere, provided that the user has adequate storage expertise. In this study, we accomplished the accurate skeletal reconstruction after resection in a virtual iliac bone tumor model in beagle dogs by combining BMP with a 3D printer, CAD, CAM, and a computer navigation system. In future, skeletal reconstruction will become possible for more specialized skeletal tissue, such as the vertebral body and load-bearing sites.

In wide resection surgery of malignant musculoskeletal tumors, optimal tumor margins and resection margin of the bone and soft tissue would be decided preoperatively on the basis of CT and/or MRI to avoid exposing the tumor tissue to the surgical field and later local recurrence of the tumor. But the accurate resection as planned preoperatively has often proven difficult. In this experimental study, a CAS system of CT-based navigation was used as a guide for bone resection margin, and the results indicated that it is possible to perform bone resection exactly as planned using the CAS system. But planning of the exact resection margin of the soft tissue covering the tumor was not attempted in this series, and similar studies based on CT and/or MRI will be necessary to determine optimal resection margin of the soft tissue.

The 3D configuration of the defect on the left iliac bone after resection of the virtual tumor was depicted on screen by subtraction of the 3D image of the left iliac bone from the 3D mirror image of the right iliac bone. CT data of the defect images were then transferred to the CAM system to drive the system to fabricate an IP-CHA implant to fill and fit the bone defect generated in the left iliac bone. The shape and size of the implants thus fabricated were acceptable and matched the defects well. However, the IP-CHA implant alone did not repair the bone defect with a new bone, as observed in control animals, in spite of its high osteoinductive capacity due to its fully interconnected porous structure. To overcome this disadvantage of the IP-CHA implant and to enhance regenerative potential of the implant, rhBMP-2 with its potent bone-inducing capacity in combination with a specific delivery system (carrier material for BMP-2) was added to the surface of the IP-CHA implant. This successfully elicited new bone formation on the surface of the implant and restored normal anatomical surface contour of the iliac bone in 12 weeks using the IP-CHA non-degradable block within the new bone mass. Replacement of the non-biodegradable IP-CHA block by a porous β-TCP implant might have resulted in complete regeneration of the bone defect without remnant biomaterials.

http://dx.doi.org/10.5772/63743

However, because of the highly fragile nature of the porous TCP and easy breakage of the implant during fabricating and surgical procedures, the IP-CHA implant was used in this study, and the bone mass with stable connection to the original iliac bone and porous implant was created.

Recent developments in 3D printing have expanded its range of application to include production of implants, scaffolds for tissue engineering, and instruments for DDSs in addition to visualization of surgical sites preoperatively and casting of surgical tools on the basis of the additive manufacturing technology used in the present study. In addition, it has become possible to directly print biological materials onto solvent-free, aqueous-based 3D scaffolds for tissue transplantation [36–38]. Using CAD/CAM techniques, spatially defined implants can be constructed by attaching functional living cells, physiological substances, and bioactive factors in layers to organ scaffolds created on the basis of MRI or CT. For this bioprinting technique, three major modes are currently available: inkjet, micro-extrusion, and laser-assisted bioprinting. These modes differ in various properties, such as the range of material viscosity accommodated, gelling method, viability, and concentration of cells to be handled. Skin and cartilage regeneration by the inkjet method, creating heart valves and blood vessels by the micro-extrusion method, and creating skin containing the heart tissue or cells by the laser-assisted method have been reported [37, 38]. We expect that for cases involving extensive tissue defects or requiring organ transplantation, printing technologies, scaffold development, cell surgical techniques, and the use of growth factors will advance further and the scope of clinical application will be further broadened if necessary tissues can be created in a shorter period of time in the future.

5. Conclusion

Advances in computer-assisted techniques have led to computer-assisted preoperative planning, custom production of surgical implants using patient data, and use of navigation systems in orthopedics. In this study, we accomplished accurate skeletal reconstruction after resection in a virtual iliac bone tumor model in beagle dogs by combining BMP with a 3D printer, CAD, CAM, and a computer navigation system. In future, skeletal reconstruction will become possible for more specialized skeletal tissues, such as the vertebral body and load-bearing sites. Moreover, in future, the development of bioprinting techniques may enable us to reconstruct extensive tissue defects or organ transplantation.

Author details

Koichi Yano[1*], Takashi Namikawa[2], Takuya Uemura[3], Yasunori Kaneshiro[1] and Kunio Takaoka[4]

*Address all correspondence to: koichiyano@hotmail.com

1 Department of Orthopaedic Surgery, Seikeikai Hospital, Sakai city, Japan

2 Department of Orthopaedic Surgery, Osaka City General Hospital, Osaka city, Japan

3 Department of Orthopaedic Surgery, Osaka City University Graduate School of Medicine, Osaka city, Japan

4 Department of Orthopaedic Surgery, Nishinomiya Watanabe Hospital, Nishinomiya city, Japan

References

[1] Buck BE, Malinin TI. Human bone and tissue allografts. Preparation and safety. Clin Orthop Relat Res. 1994;Jun(303):8–17.

[2] Arrington ED, Smith WJ, Chambers HG, Bucknell AL, Davino NA. Complications of iliac crest bone graft harvesting. Clin Orthop Relat Res. 1996;Aug(329):300–9.

[3] Butler D. Last chance to stop and think on risks of xenotransplants. Nature. 1998;22(391):320–4.

[4] DeLustro F, Dasch J, Keefe J, Ellingsworth L. Immune responses to allogeneic and xenogeneic implants of collagen and collagen derivatives. Clin Orthop Relat Res. 1990;Nov(260):263–79.

[5] Oryan A, Alidadi S, Moshiri A, Maffulli N. Bone regenerative medicine: classic options, novel strategies, and future directions. J Orthop Surg Res. 2014;17(9):24628910. DOI: 10.1186/1749-799X-9-18

[6] Boden SD, Kang J, Sandhu H, Heller JG. Use of recombinant human bone morphogenetic protein-2 to achieve posterolateral lumbar spine fusion in humans: a prospective, randomized clinical pilot trial: 2002 Volvo Award in clinical studies. Spine (Phila Pa 1976). 2002;27(23):2662–73.

[7] Burkus JK, Transfeldt EE, Kitchel SH, Watkins RG, Balderston RA. Clinical and radiographic outcomes of anterior lumbar interbody fusion using recombinant human bone morphogenetic protein-2. Spine (Phila Pa 1976). 2002;27(21):2396–408.

[8] Friedlaender GE, Perry CR, Cole JD, Cook SD, Cierny G, Muschler GF, Zych GA, Calhoun JH, LaForte AJ, Yin S. Osteogenic protein-1 (bone morphogenetic protein-7) in the treatment of tibial nonunions. J Bone Joint Surg Am. 2001;83(Suppl 1):S151–8.

[9] Brown GA, Milner B, Firoozbakhsh K. Application of computer-generated stereolithography and interpositioning template in acetabular fractures: a report of eight cases. J Orthop Trauma. 2002;16(5):347–52.

[10] Takeyasu Y, Oka K, Miyake J, Kataoka T, Moritomo H, Murase T. Preoperative, computer simulation-based, three-dimensional corrective osteotomy for cubitus varus

deformity with use of a custom-designed surgical device. J Bone Joint Surg Am. 2013;95(22):e173. DOI: 10.2106/JBJS.L.01622

[11] Bellemans J, Vandenneucker H, Vanlauwe J. Robot-assisted total knee arthroplasty. Clin Orthop Relat Res. 2007;Nov(464):111–6.

[12] Fujishiro T, Nakaya Y, Fukumoto S, Adachi S, Nakano A, Fujiwara K, Baba I, Neo M. Accuracy of pedicle screw placement with robotic guidance system: a cadaveric study. Spine (Phila Pa 1976). 2015;40(24):1882–9. DOI: 10.1097/BRS.0000000000001099

[13] Ieguchi M, Hoshi M, Takada J, Hidaka N, Nakamura H. Navigation-assisted surgery for bone and soft tissue tumors with bony extension. Clin Orthop Relat Res. 2012;470(1): 275–83. DOI: 10.1007/s11999-011-2094-5

[14] Yoshikawa H, Tamai N, Murase T, Myoui A. Interconnected porous hydroxyapatite ceramics for bone tissue engineering. J R Soc Interface. 2009;6(Suppl 3):S341–8. DOI: 10.1098/rsif.2008.0425

[15] Saito N, Okada T, Horiuchi H, Murakami N, Takahashi J, Nawata M, Ota H, Nozaki K, Takaoka K. A biodegradable polymer as a cytokine delivery system for inducing bone formation. Nat Biotechnol. 2001;19(4):332–5.

[16] Miyamoto S, Takaoka K, Okada T, Yoshikawa H, Hashimoto J, Suzuki S, Ono K. Polylactic acid-polyethylene glycol block copolymer. A new biodegradable synthetic carrier for bone morphogenetic protein. Clin Orthop Relat Res. 1993;Sep(294):333–43.

[17] Saito N, Okada T, Toba S, Miyamoto S, Takaoka K. New synthetic absorbable polymers as BMP carriers: plastic properties of poly-D,L-lactic acid-polyethylene glycol block copolymers. J Biomed Mater Res. 1999;47(1):104–10.

[18] Saito N, Okada T, Horiuchi H, Murakami N, Takahashi J, Nawata M, Ota H, Miyamoto S, Nozaki K, Takaoka K. Biodegradable poly-D,L-lactic acid-polyethylene glycol block copolymers as a BMP delivery system for inducing bone. J Bone Joint Surg Am. 2001;83-A Suppl 1(Pt 2):S92–8.

[19] Murakami N, Saito N, Horiuchi H, Okada T, Nozaki K, Takaoka K. Repair of segmental defects in rabbit humeri with titanium fiber mesh cylinders containing recombinant human bone morphogenetic protein-2 (rhBMP-2) and a synthetic polymer. J Biomed Mater Res. 2002;62(2):169–74.

[20] Murakami N, Saito N, Takahashi J, Ota H, Horiuchi H, Nawata M, Okada T, Nozaki K, Takaoka K. Repair of a proximal femoral bone defect in dogs using a porous surfaced prosthesis in combination with recombinant BMP-2 and a synthetic polymer carrier. Biomaterials. 2003;24(13):2153–9.

[21] Saito N, Okada T, Horiuchi H, Ota H, Takahashi J, Murakami N, Nawata M, Kojima S, Nozaki K, Takaoka K. Local bone formation by injection of recombinant human bone morphogenetic protein-2 contained in polymer carriers. Bone. 2003;32(4):381–6.

[22] Yoneda M, Terai H, Imai Y, Okada T, Nozaki K, Inoue H, Miyamoto S, Takaoka K. Repair of an intercalated long bone defect with a synthetic biodegradable bone-inducing implant. Biomaterials. 2005;26(25):5145–52.

[23] Kato M, Toyoda H, Namikawa T, Hoshino M, Terai H, Miyamoto S, Takaoka K. Optimized use of a biodegradable polymer as a carrier material for the local delivery of recombinant human bone morphogenetic protein-2 (rhBMP-2). Biomaterials. 2006;27(9):2035–41.

[24] Suzuki A, Terai H, Toyoda H, Namikawa T, Yokota Y, Tsunoda T, Takaoka K. A biodegradable delivery system for antibiotics and recombinant human bone morphogenetic protein-2: a potential treatment for infected bone defect. J Orthop R. 2003;24(3): 327–32.

[25] Saito N, Murakami N, Takahashi J, Horiuchi H, Ota H, Kato H, Okada T, Nozaki K, Takaoka K. Synthetic biodegradable polymers as drug delivery systems for bone morphogenetic proteins. Adv Drug Deliv Rev. 2005;57(7):1037–48.

[26] Kato M, Namikawa T, Terai H, Hoshino M, Miyamoto S, Takaoka K. Ectopic bone formation in mice associated with a lactic acid/dioxanone/ethylene glycol copolymer-tricalcium phosphate composite with added recombinant human bone morphogenetic protein-2. Biomaterials. 2006;27(21):3927–33.

[27] Namikawa T, Terai H, Suzuki E, Hoshino M, Toyoda H, Nakamura H, Miyamoto S, Takahashi N, Ninomiya T, Takaoka K. Experimental spinal fusion with recombinant human bone morphogenetic protein-2 delivered by a synthetic polymer and beta-tricalcium phosphate in a rabbit model. Spine (Phila Pa 1976). 2005;30(15):1717–22.

[28] Hoshino M, Egi T, Terai H, Namikawa T, Takaoka K. Repair of long intercalated rib defects using porous beta-tricalcium phosphate cylinders containing recombinant human bone morphogenetic protein-2 in dogs. Biomaterials. 2006;27(28):4934–40.

[29] Hoshino M, Namikawa T, Kato M, Terai H, Taguchi S, Takaoka K. Repair of bone defects in revision hip arthroplasty by implantation of a new bone-inducing material comprised of recombinant human BMP-2, Beta-TCP powder, and a biodegradable polymer: an experimental study in dog. J Orthop R. 2007;25(8):1042–51.

[30] Taguchi S, Namikawa T, Ieguchi M, Takaoka K. Reconstruction of bone defects using rhBMP-2-coated devitalized bone. Clin Orthop Relat Res. 2007;Aug(461):162–9.

[31] Tokuhara Y, Wakitani S, Imai Y, Kawaguchi A, Fukunaga K, Kim M, Kadoya Y, Takaoka K. Repair of experimentally induced large osteochondral defects in rabbit knee with various concentrations of Escherichia coli-derived recombinant human bone morphogenetic protein-2. Int Orthop. 2010;34(5):761–7.

[32] Hoshino M, Egi T, Terai H, Namikawa T, Kato M, Hashimoto Y, Takaoka K. Repair of long intercalated rib defects in dogs using recombinant human bone morphogenetic

protein-2 delivered by a synthetic polymer and beta-tricalcium phosphate. J Biomed Mater Res A. 2009;90(2):514–21.

[33] Yano K, Hoshino M, Ohta Y, Manaka T, Naka Y, Imai Y, Sebald W, Takaoka K. Osteoinductive capacity and heat stability of recombinant human bone morphogenetic protein-2 produced by *Escherichia coli* and dimerized by biochemical processing. J Bone Miner Metab. 2009;27(3):355–63. DOI: 10.1007/s00774-009-0040-3

[34] Li J, Li Y, Ma S, Gao Y, Zuo Y, Hu J. Enhancement of bone formation by BMP-7 transduced MSCs on biomimetic nano-hydroxyapatite/polyamide composite scaffolds in repair of mandibular defects. J Biomed Mater Res A. 2010;95(4):973–81. DOI: 10.1002/jbm.a.32926

[35] Nakamura A, Akahane M, Shigematsu H, Tadokoro M, Morita Y, Ohgushi H, Dohi Y, Imamura T, Tanaka Y. Cell sheet transplantation of cultured mesenchymal stem cells enhances bone formation in a rat nonunion model. Bone. 2010;46(2):418–24. DOI: 10.1016/j.bone.2009.08.048

[36] Do AV, Khorsand B, Geary SM, Salem AK. 3D printing of scaffolds for tissue regeneration applications. Adv Healthc Mater. 2015;4(12):1742–62. DOI: 10.1002/adhm.201500168

[37] Chia HN, Wu BM. Recent advances in 3D printing of biomaterials. J Biol Eng. 2015;9(4):25866560. DOI: 10.1186/s13036-015-0001-4

[38] Murphy SV, Atala A. 3D bioprinting of tissues and organs. Nat Biotechnol. 2014;32(8):773–85. DOI: 10.1038/nbt.2958

Section 3

Perspective Powder Additive Manufacturing

Chapter 8

Applications of the Selective Laser Melting Technology in the Industrial and Medical Fields

Pacurar Razvan and Pacurar Ancuta

Additional information is available at the end of the chapter

http://dx.doi.org/10.5772/63038

Abstract

The existence of an MCP Realizer II SLM 250 equipment at the National Centre of Innovative Manufacturing from the Technical University of Cluj-Napoca (TUC-N) facilitated the starting of different research activities in this field at TUC-N with the aim of improving the Selective Laser Melting (SLM) process capability for a better transfer of the technological gained knowledge to different partners from the industrial and medical fields. Reaching this goal has been also facilitated by the fact that in the period 2010–2013 at the Technical University of Cluj-Napoca, within a postdoctoral project financed by European Commission (E.U.), it was possible to activate with the program entitled "Research regarding the manufacturing of metallic parts by Selective Laser Melting (SLM) technology." Part of the results obtained in this postdoctoral program are presented within this chapter of the book, being addressed not only to the engineers, PhD students, researchers, medical doctors, but also to anyone who might be interested about this Additive Manufacturing (AM) method and its possible applications in the industrial and medical fields.

Keywords: selective laser melting, SLM, additive manufacturing, tool steel, H13, customized medical implants, titanium, TiAl6V4, finite element analysis, FEA

1. Introduction

Nowadays, the Selective Laser Melting (SLM) technology is widely used in different domains of the industry, such as aerospace, automotive, consumer goods and medical field. This additive manufacturing technology method offers an important series of advantages as compared to the conventional manufacturing methods that consists in the capability of realizing different parts with high complexity of the shape, the reduced manufacturing time, etc. By using the SLM method, it is possible to manufacture finite parts without necessarily using other supplementary manufacturing processes. The mechanical properties of the realized parts are acceptable, being dependent on the composition and the size of the metallic powder grains, and also on the internal structure of the parts, process parameters, and the manufacturing strategy used. These factors also have an important influence on the surface roughness and the accuracy of the manufactured parts.

The working principle of the SLM process is easy to understood, as could be observed in **Figure 1**. There are several steps that have to be fulfilled. Before starting the process, an inert gas (nitrogen or argon) will be introduced in the SLM machine chamber through a circulating system as presented in **Figure 1**, until an inert atmosphere will be obtained inside the working chamber (the maximum level of oxygen admitted inside is 0.1%). Furthermore, the inert gas will be circulated through the system during the entire process until the end with a level of 0.1% of oxygen maintained constant.

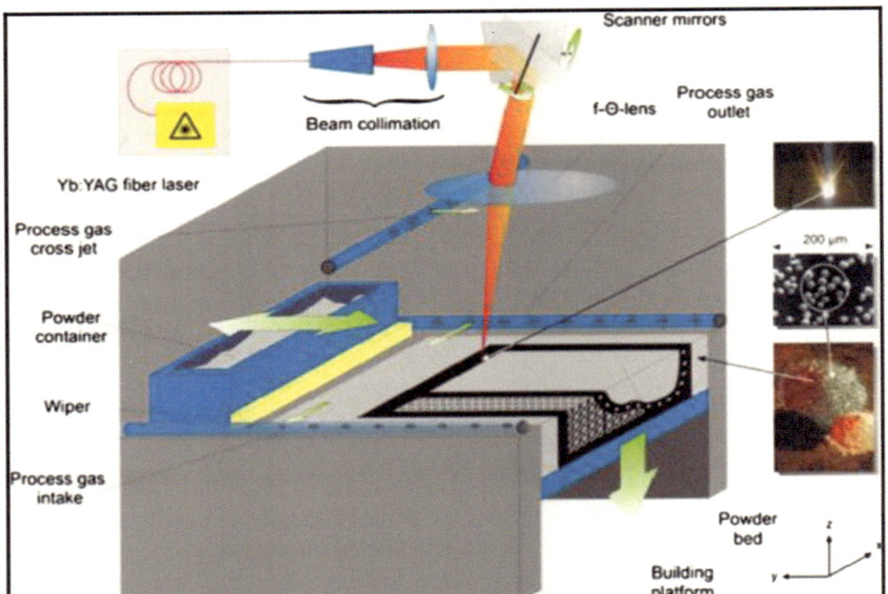

Figure 1. Working principle of the SLM process [1].

The process continues with the deposition of the raw material from the powder container over the building platform by using a wiper that moves along the Y-axis direction. The thickness of the deposited material is around 20–100 μm for the first few layers. The process is repeated until the powder covers the entire building platform uniformly. Then the layer thickness will be set to a constant value of 30–50 μm and the process continues with the scanning of the first layer according to the first slice of the model, and then this will continue with the scanning of the next slice and so on until the part will be finished on the machine.

The building platform is moved along the Z-axis after the scanning process of every layer. Finally, when the maximum height of the building packet is reached, the building platform is moved to the start position and the manufactured packet is removed from the platform after sucking the non-melted metallic powder and after eliminating the metallic supports that were generated within the control software of the SLM system. The metallic supports have a wired structure and are needed in the manufacturing process in order to maintain the built part onto the manufacturing platform.

The SLM technology looks quite simple in theory, but the process control is not so simple from the experimental point of view. Experts' opinions regarding the technological parameters that have a significant influence onto the accuracy and the mechanical properties of a manufactured part made by SLM are quite different [1]. There are researchers who state that these characteristics are directly influenced by the laser system. So, the laser power or the laser beam diameter is important when speaking about the accuracy or porosity of a manufactured part [2, 3]. There are other researchers that state that the optical system in close connection to the scanning strategy is the most important when speaking about the accuracy or porosity of the manufactured part using the SLM equipment. The possibility of cooling-down the optics or optic's design is one of the most important issues in this case [4, 5]. Other researchers state that the scanning speed, the layer thickness or the building temperature have a direct influence over the SLM process and the accuracy or porosity of the manufactured parts as well [6–9]. There are also a series of researchers which consider that the properties of the raw material (particle size, particle distribution, etc.) are important within the SLM process, especially in the case when the resulted porosity of the material has to be precisely controlled [10–11]. Other researchers state that innovative technologies should be used in combination with the classical manufacturing technologies in the so called "hybrid technologies" [12–15]. This means, for example, that the parts have to be manufactured by using the SLM technology and finishing of the part can be done on the same equipment by using conventional manufacturing methods, such as milling. In this way, the accuracy of the manufactured part and the surface roughness will be improved significantly. Looking on all these various researches reported by the authors, it is very difficult to state who is right and who is not right regarding the best control of the SLM process.

The existence of an MCP Realizer II SLM 250 equipment (see **Figure 2**) at the National Centre of Innovative Manufacturing from the Technical University of Cluj-Napoca (TUC-N) facilitated the start of different research activities in this field at TUC-N with the aim of improving the Selective Laser Melting (SLM) process capability for a better transfer of the technological gained knowledge to different partners from the industrial and medical fields. Reaching this goal has been also facilitated by the fact that in the period 2010–2013 at the Technical University of Cluj-Napoca, within a postdoctoral project financed by European Commission (E.U.), it was possible to activate with the program entitled "Research regarding the manufacturing of metallic parts by Selective Laser Melting (SLM) technology."

Figure 2. MCP Realizer II SLM 250 system from the Technical University of Cluj-Napoca.

Part of the results obtained in this postdoctoral program, in cooperation with the SLM Solutions GmbH Company from Luebeck, Germany (best practice examples) are presented in this chapter of the book.

2. Injection moulding tools made from H13 material using the selective laser melting technology (industrial case study)

One of the most important applications of the Selective Laser Melting (SLM) technology is the manufacturing of tools for the injection moulding process. Besides the advantages of the SLM process that consists in the fact that it is rapid, does not depend on the shape of the tool and it has reasonable costs as compared to the classical manufacturing technologies, there are some disadvantages as well. One of the major disadvantages of the SLM process is the accuracy, which is below the accuracy provided by some well-known classical technologies, such as grinding. The SLM technology looks quite simple in theory, but the process control is not so simple from the experimental point of view. The research started within the Technical University of Cluj-Napoca, in cooperation with Plastor SA Company from Oradea (Romania) and SLM Solutions GmbH Company from Luebeck (Germany) has proved the complexity of the SLM technology in the manufacturing of the injection moulding tools process. There are many important aspects that have to be taken into account, as mentioned above, when speaking about the accuracy of the manufactured injection moulding tools, in close-connection with the accuracy of the SLM process.

2.1. Finite element analysis used to estimate the shrinkage of the SLM tools

The case study to be analyzed refers to the injection moulding tool components (punch and die) for a lid button of a grass-cutting machine manufactured by Plastor SA Company from Oradea (Romania) (see **Figure 3**).

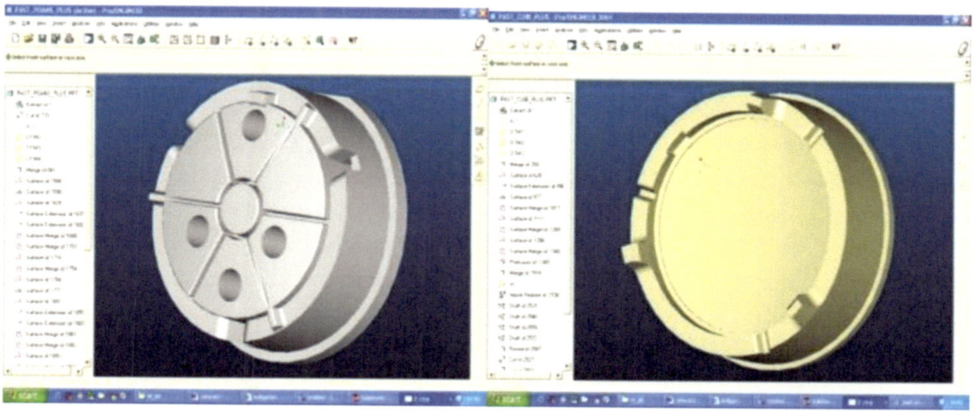

Figure 3. Injection moulding tools for a lid button (ProEngineer design).

The 3D CAD models were imported afterward into the Ansys FEA program and the mesh was generated as illustrated in **Figure 4**. A mesh with a total number of 31.930 nodes and 18.618 elements for the punch and 28.276 nodes and 15.118 elements for the die, respectively, were generated within the FEA program.

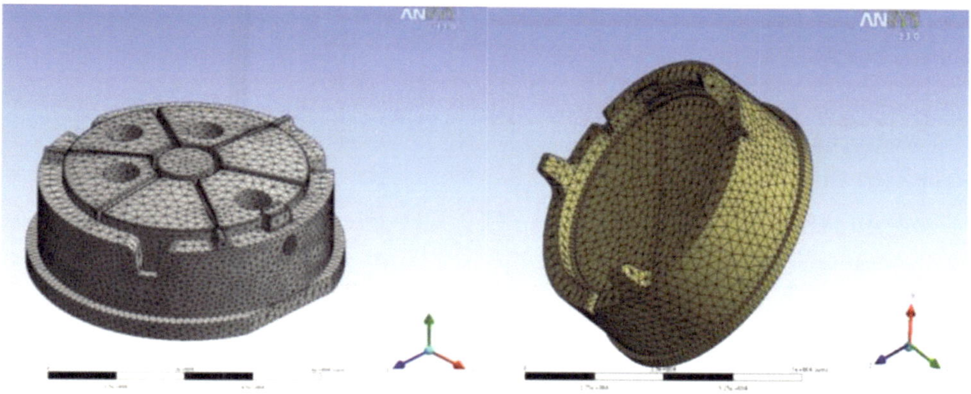

Figure 4. The mesh generated for the injection moulding tools—punch and die (Ansys 13).

Another important aspect consists in the introduction of thermal and mechanical characteristics of the raw material, as it was specified by the producer of the type of metallic powder that is commercially available for such applications: H13 Tool Steel material (see **Table 1**) [16–17].

The injection moulding tools were manufactured using the SLM 250 HL equipment from SLM Solutions GmbH Company in Luebeck, Germany from this type of material, having the same characteristics as the ones that were introduced within the finite element analysis (H13 Tool Steel material).

The next step of the finite element analysis consisted in the introduction of technological parameters within the Ansys program. The laser power, scanning speed and powder bed temperature were introduced as a subroutine into the analysis within the ANSYS APDL Heat Transfer Module.

Twenty finite element analyses were done using varied values of the technological parameters specified in **Table 2**, in order to find the combination of parameters that leads to an optimum closing of the active elements of moulds (see **Figure 5** and **Table 3**).

Property	Value
Density	7.80 (g/cm^3)
Rockwell C hardness	54 HRC
Tensile strength	1730 MPa
Elongation	10.0%
Specific heat at 20°C	460 (J/kg K)
Thermal conductivity	28.6 (W/mK)
Melting point	1020°C
Poisson coefficient	0.3
Elastic modulus	200 GPa

Table 1. H13 Tool Steel material properties.

Technological parameter	Varied value					
Laser power [W]	175	180	185	190	195	200
Scanning speed [mm/s]	250	300	350	400	450	500
Powder bed temperature [°C]	80	104	128	152	176	200

Table 2. Technological parameters introduced in analysis using the APDL Heat transfer module.

Besides these technological parameters that were varied within the finite element analysis, there were also other important parameters that were specified, but were maintained constant during the analyses, such as the layer thickness (30 μm) and the hatching distance (20 μm). As it is possible to observe from the analysis, the value of technological parameters that leads to a minimum shrinkage in the case of punch and die made by SLM are different. If the temperature of the powder bed has the same value in both cases (176°C), the laser power and the scanning speed should be different, such as at the end the punch and die would fit by the closing point of view.

Item	Laser power [W]	Scanning speed [mm/s]	Powder bed temperature [°C]	Shrinkage Δx [μm]	Shrinkage Δy [μm]
Punch	175	300	176	70.162	66.439
Die	200	400	176	71.496	73.231

Table 3. Results of the finite element analyses.

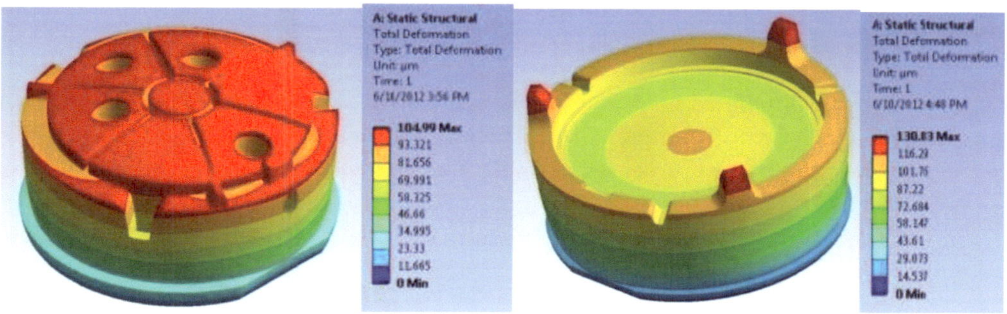

Figure 5. The resulted shrinkage of the punch and die.

One may conclude that in the case of SLM technology it is very difficult to find a set of technological parameters that would be unique and universally valid in all cases of moulds manufactured by SLM from H13 Tool Steel material. The technological parameters to be used within the SLM process have to be different, according to the type and the shape of the metallic tools to be manufactured. The laser power and the energy density being applied on the scanned surface are obviously different if the injection moulding tools have thin walls or these tools are solid and massive in the entire structure of the material.

2.2. Injection moulding tools made by SLM equipment and measurements

Besides setting up the technological parameters, there is another important aspect regarding the accuracy issue, such as the support generating stage that is actually a pre-processing stage needed in the SLM process, in order to sustain the manufactured part onto the building platform of the machine.

The metallic supports have a wired structure and are needed because of the high stresses that occur during the welding process, having the tendency of severely deforming the manufactured models during manufacturing process, especially if welded connection is not good on the building platform starting with the first layers of the manufactured parts while manufacturing. The metallic supports were generated, for the punch and die, using the Magics 15 software, as one may observe in **Figure 6**. The metallic supports are removed within the post-processing stage.

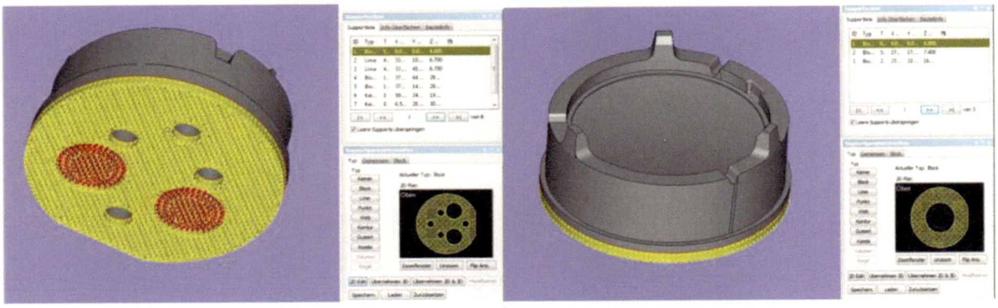

Figure 6. Supports generated for the punch and die using Magics 15 software.

The punch and die illustrated in **Figure 7** were manufactured on the SLM 250 HL equipment from SLM Solutions GmbH Company from Luebeck (Germany) presented in **Figure 8**, using the technological parameters presented in **Table 3**, as they were determined within the finite element analyses.

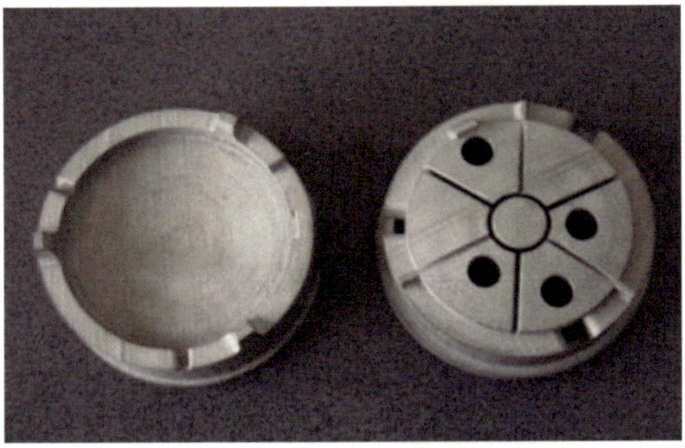

Figure 7. The injection moulding tools made by SLM equipment.

Some measurements of the injection moulding tools were also made at the Technical University of Cluj-Napoca using a Zeiss Eclipse 550 CMM equipment. The conclusion was that the obtained results are comparable to the ones estimated within the finite element analysis (designed dimensions). The maximum value of shrinkage that has been experimentally determined has a value of approximately 80 μm, both in the case of the punch and die (see **Figures 9** and **10**).

The lowest values of deformations (less than 10 μm) were obtained in the case of the dimensions H1, used for the correct positioning of the punch and die. The mean value of the punch and die shrinkage has been determined in the interval of 30–40 μm, values that are comparable to the ones obtained in the case when the punch and die are produced by using similar technologies dealing with metallic powders such as Selective Laser Sintering (SLS) and Classical Sintering (CS).

Figure 8. The SLM 250 HL equipment at SLM Solutions GmbH Company from Luebeck, Germany.

There are still other issues to be investigated in the near future such as finding a way for a better control of the SLM process.

The errors that occur during the SLM process can be compensated if precisely calculated scale factors would be applied in the pre-processing stage onto the 3D model that has to be manufactured using the SLM equipment.

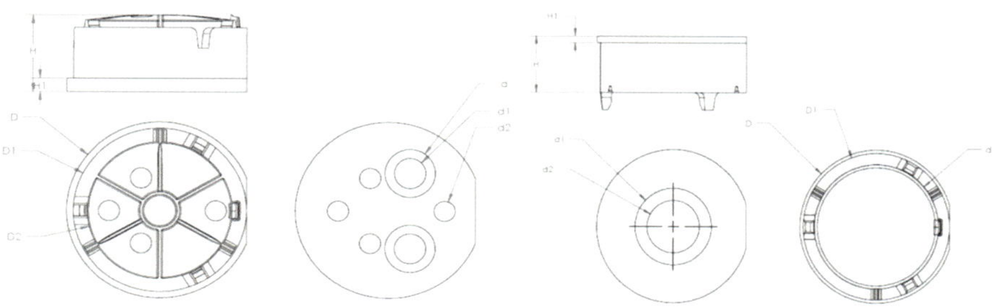

Figure 9. Schematic draw of the dimensions measured in case of punch and die.

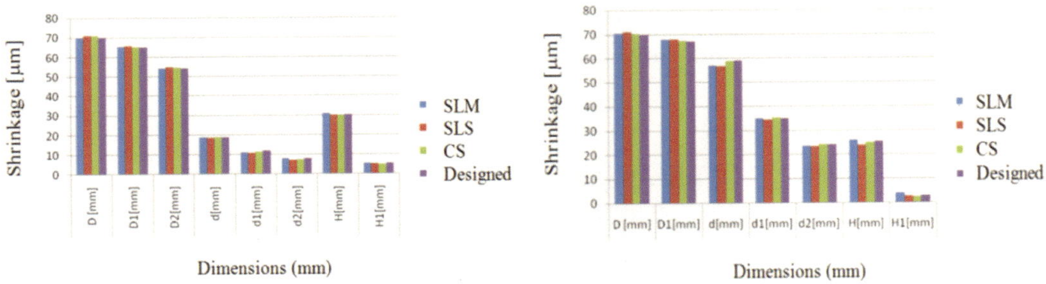

Figure 10. Shrinkage values in microns determined in the case of punch and die.

2.3. Testing the active element of moulds made by SLM at S.C. Plastor S.A. Oradea (injection moulding company from Romania)

In order to test the active elements of the mould made by SLM technology within the Plastor S.A. company from Oradea, it was required the manufacturing of additional fixing plates using conventional technologies at Plastor SA, as it is possible to observe in **Figure 11**. These plates were mounted onto the injection moulding machine Arburg 370 CMD 800-325 - type that is available within Plastor SA Company from Oradea.

Before the injection moulding experiment were made, tests were required to be done using different finite element analyses through the Autodesk Simulation Moldflow Adviser program, in order to determine the plastic injection parameters, such as the injection pressure, filling speed, etc. Four types of plastic materials were tested, such as Acryl Butadiene Styrene – ABS, Polypropylene – PP, Polyamide armed with glass fibers - PA+30% GF and Poly Oxy Methylene– POM.

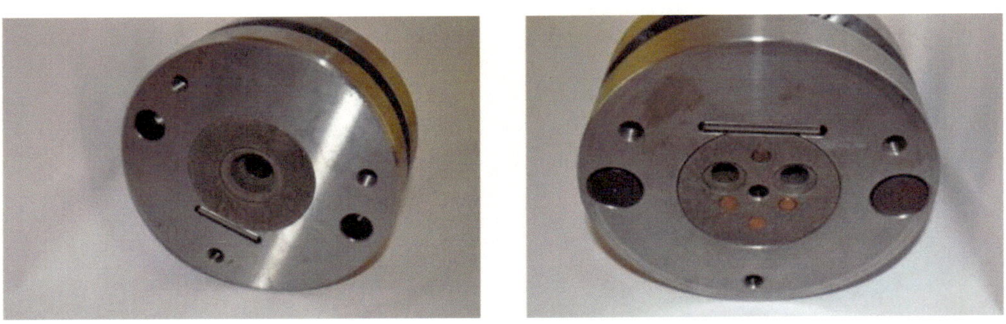

Figure 11. Active elements of moulds made by SLM mounted onto the plates manufactured at Plastor S.A. company from Oradea.

Figure 12. Arburg 370 CMD 800-325 injection moulding equipment (Plastor SA Company Oradea).

After the mesh was generated and the optimum injection point was specified, as one could be observed in **Figure 13**, in accordance with the fiber plastic material orientation, the next step consists in determining of the injection moulding technological parameters, such as the filling speed, injection pressure, cooling time, etc.

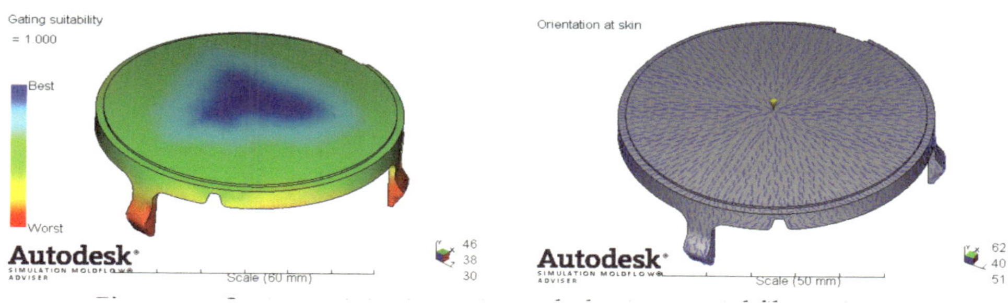

Figure 13. Optimum injection point and plastic material fiber orientation.

Figure 14 presents the values obtained for the injection pressure and the filling time in the case of ROTEC® ABS 1001 FR V0/4 material. As it is possible to observe in these images, the injection pressure required in this case is 5.130 MPa and the filling time is 0.7437 seconds.

Regarding the other plastic injection technological parameters (filling speed, melting temperature, cooling time), it is important to specify the fact that these parameters have a significant importance in the plastic injection process. Filling speed depends on the injection pressure and the properties of the plastic material. The cooling time is directly correlated with the capacity of injection moulding tool to conduct the thermal energy at the end.

An important aspect regarding the finite element analysis that has been performed is represented by the volumetric shrinkage at the end of the injection moulding process. As it is possible to observe in **Figure 15**, the volumetric shrinkage in the case of ABS plastic material were less than 4% for more than 90% of total part surface. In the clamping areas, the volumetric shrinkage values were higher, being closed to a value of 7%.

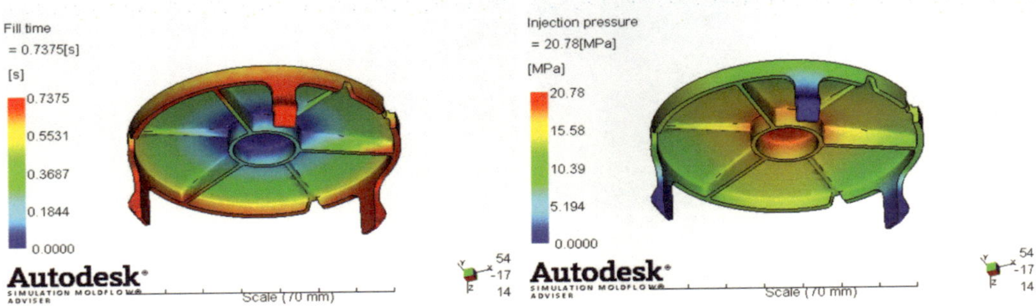

Figure 14. Injection pressure and filling time in the case of ABS plastic material.

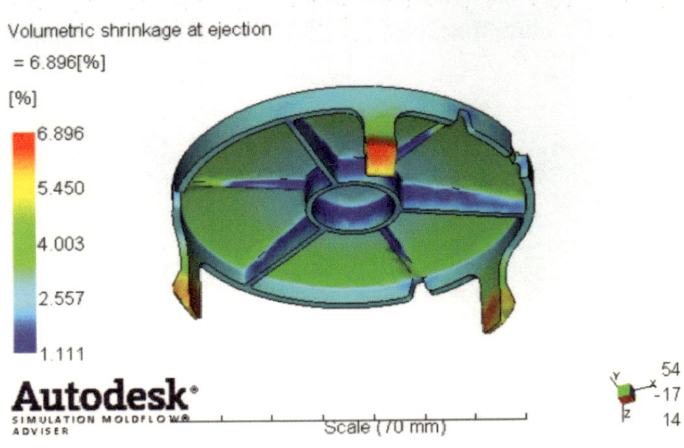

Figure 15. Volumetric shrinkage at ejection at the end of injection moulding process.

The results obtained using the finite element analyses made using the MoldFlow program in the case of all plastic materials – Acryl Butadiene Styrene – ABS, Polypropylene – PP, Polyamide armed with glass fibers - PA+30% GF and Poly Oxy Methylene– POM- are presented in **Table 4**.

The technological parameters presented in **Table 4** were used for the injection moulding tests that were made at Plastor SA Company from Oradea. All plastic materials were dried in an oven for three hours at a temperature of 80°C, with the exception of the Poly-propylene (PP)

http://dx.doi.org/10.5772/63038

material which is not required to be dried. The maximum clamping force used during the injection moulding tests was 5 T for all four types of plastic materials that were tested.

Type of material / Parameter	PP	ABS	PA	POM
- Injection pressure [MPa]	5.13	20.78	10.32	10.39
- Filling time [s]	0.74	0.74	0.84	3.7
- Cooling time [s]	39.73	26.48	18.33	63.91
- Injection temperature [°C]	230	220	280	200
- Volumetric shrinkage [%]	11.52	6.89	11.76	14.18
- Clamping force[T]	0.99	3.8	2.02	2.01

Table 4. Results obtained using finite element analyses made by using Mold Flow program.

Figure 16. Injection moulding tools mounted on the Arburg 370 CMD 800-325 testing machine from Plastor SA Oradea.

The injection pressure is different in accordance with the type of the plastic material that was tested. The injection pressure was for example 50 bars in the case of PP material and 210 bars in the case of ABS material. This fact was mainly caused by the fact that the PP material had a density of 0.9 g/cm^3 as compared to the ABS material which had a density of 1.2 g/cm^3. In order to fill in the cavity with plastic material in the second case, a pressure difference of 160 bars is necessary.

The cooling time was adjusted differently in accordance with the plastic material that was injected into the mould. In the case of PA plastic material, a total time of 15 seconds is required, as compared with the ABS plastic material where the cooling time required is 50 seconds. This can be explained by the fact that the thermal conductivity coefficient is different in these cases. In the case of PA material, the thermal conductivity coefficient is 0.36 W/mK, while in the case of the ABS plastic material the value of this coefficient is 0.17 W/mK. The difference of 0.19 W/mK is transformed to the time difference required to cool-down the material in a supplementary period of 35 seconds.

2.4. Conclusion

As a conclusion of made research, it is possible to state that the Selective Laser Melting technology is easy to be understood in principle, but it is not so easy to be controlled. There are a lot of aspects that have to be taken into consideration when speaking about the accuracy of the injection moulding tools made by Selective Laser Melting (SLM), starting with the properties of the raw material, the optical system and ending with the scanning strategy or the technological parameters that are used in the manufacturing process. As related to the technological parameters (laser power, scanning speed, powder bed temperature, etc.), as it has been proven by the finite element analyses that were made, it is very difficult to find a set of technological parameters that would be unique and universally valid in all cases of moulds manufactured by Selective Laser Melting from H13 Tool Steel material. The accuracy of the injection-moulded tools made by Selective Laser Melting technology will be different, being dependent on the geometry of the tools (the size and the shape) and the accuracy of the process. Research still needs to be done in the future regarding the determination of scale factors that can be applied in the pre-processing stage onto the 3D model that has to be manufactured using the Selective Laser Melting equipment. The injection moulding tools were successfully manufactured on an SLM 250 HL equipment in the SLM Solutions GmbH Company from Luebeck (Germany) and tested in the injection moulding process of four type of plastic materials (Acryl Butadiene Styrene – ABS, Polypropylene – PP, Polyamide armed with glass fibers - PA+30% GF and Poly Oxy Methylene– POM) at Plastor SA Company from Oradea (Romania), as it is possible to observe in **Figure 17**.

Figure 17. Injected plastic materials obtained at Plastor SA Oradea using the injection moulding tools made by SLM.

3. Medical implants with lattice structures made from Titanium by using the Selective Laser Melting Technology (medical case study example)

3.1. Introduction

The recent researches developed in the bio-medical field proved the high interest that exists in this field regarding the possibilities of manufacturing customized medical implants using Additive Manufacturing (AM) methods, such as Selective Laser Sintering (SLS), Selective Laser Melting (SLM) or Electron Beam Melting (EBM) [18–22]. These types of AM methods allow the manufacturing of fully dense metallic parts with complex geometrical shape, starting from a 3D model realized with a computer aided design (CAD) program and exported in an "*.stl" format [23]. Fully dense metallic structures are required to be produced in the case of customized medical implants made by SLM, especially in some regions where the implants needs to be fixed into the human bone with titanium screws. There are also regions of the implants where the porous structure is required for a better osseointegration process of the human tissue through the surface of the medical implant made by SLM. Obtaining structures with a well-controlled level of porosity in this case is very important and could be achieved in different ways, such as by designing an implant having different types of lattice structures (geometrical configuration of cells) by adjusting the technological parameters (laser power, hatching distance, etc.) and the scanning strategy during the SLM manufacturing process, or by mixing in different ratios the raw powder material (titanium, in this case) with other types of biocompatible materials (e.g. hydroxyapatite, PMMA, etc.) [24]. The research presented in this chapter, made at the Technical University of Cluj-Napoca (TUC-N), was focused on the finite element analysis of the strain and stress of several models that were especially designed to have different types of lattice structures (size and geometrical configuration of cells). The samples designed and made by SLM at TUC-N using the MCP Realizer II SLM 250 equipment were analyzed afterwards by using a Scanning Electron Microscope JSM – 5600 LV (JEOL) type, in order to determine which is the optimum size and configuration of the cells to be recommended from the structural point of view to be used within the design and manufacturing process of a customized medical implant to be made by SLM. Taking into account the results obtained at TUC-N, a customized medical implant was manufactured by SLM from TiAl6V4 material for a German Medical Institute, by using the SLM 250 HL equipment from SLM Solutions GmbH Company (Luebeck, Germany), at the end.

3.2. Design of the lattice structures using the SolidWorks CAD program

Six types of models similar to the ones presented in **Figure 18** were designed using SolidWorks, with different types of lattice structures (geometrical features), as following:

- rhombic shape (size of a single cell=1 mm)
- hexagonal shape (diameter of a single cell=Φ1.45 mm)
- square shape rounded at the edges (size of a single cell: l=0.8 mm; r=0.1 mm)
- rectangular shape (size of a single cell: l_1=1 mm; l_2=0.6 mm)

- circular shape (diameter of a single cell: d=1 mm)
- pyramidal shape (base size of a single cell: l=1.5 mm).

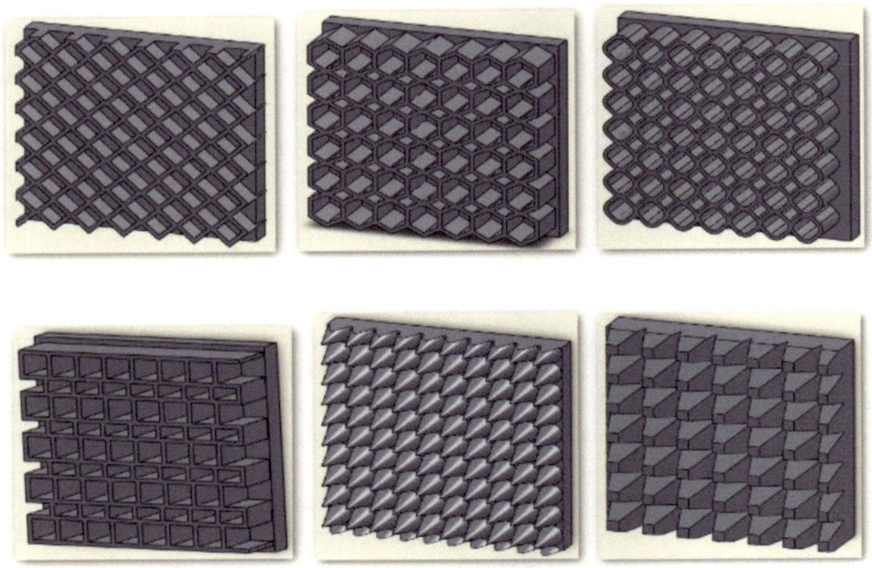

Figure 18. Models with different geometrical features designed using SolidWorks.

The extrusion and width of the cells was designed as having 0.2 mm in the case of all models that were designed using SolidWorks. The cellular structure has been obtained by copying the shape of a single cell along the X and Y-axes directions. The models designed in this way were further analyzed using the Abaqus finite element program, in order to determine the stress and strain of the samples and their mechanical behaviour, in the case of a particular pressure load that has been applied in the uniaxial direction onto the top surface of the designed models.

3.3. Estimating the mechanical behaviour of samples with different types of lattice structures manufactured by SLM

The mechanical behaviour of all models presented in **Figure 18** was analyzed using the Abaqus 6.9-3 FEA program. Several material characteristics, such as the elastic modulus (E=114 GPa), Poisson ratio (υ=0.31) and yield strength (σ_c= 775 MPa) for the TiAl6V4 metallic powder material were taken into consideration in the analyses, as they are specified in the datasheet of the material provided by the supplier company on its website. [17]

The next step consisted in establishing the movement restrictions along the X, Y, and Z-axes, as illustrated in **Figure 19**.

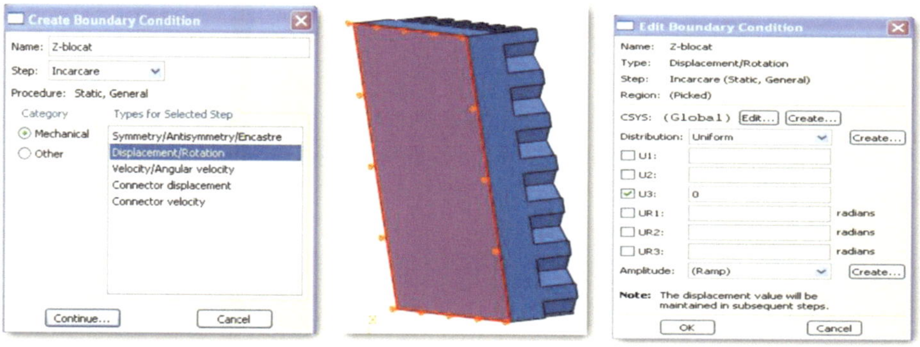

Figure 19. Movement restriction applied on the Z-axis direction.

As one may notice, the "Displacement" type of restriction has been selected in this case. The degrees of freedom were locked accordingly along all three axes by selecting the adequate facet of the model and by setting up the corresponding displacement value to zero.

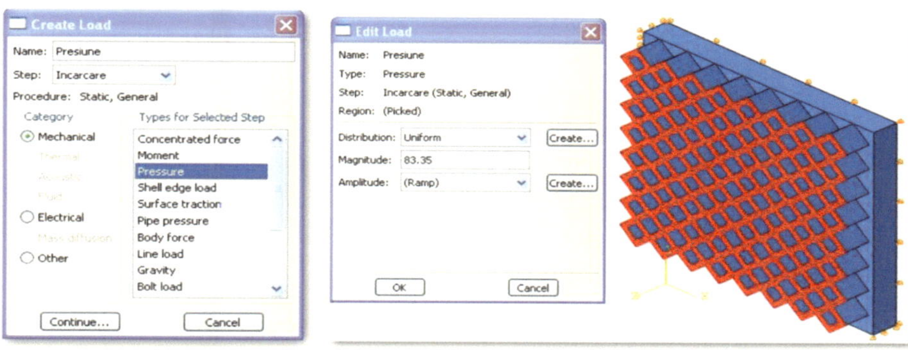

Figure 20. Applied load specified in the analysis.

The pressure applied in a single direction onto the top surface of the model was determined by using formula (1):

$$P = \frac{F}{A_p} \tag{1}$$

A force value F = 3200 N, which corresponds to four-times higher the force exerted by a person weighing 80 kg has been considered in all analyzed cases. The value of the contact area A_p in mm^2 was different from case to case, due to the fact that the geometrical configuration of the cells was different. The A_p value was determined for each case by using the tools that are available within the SolidWorks CAD program. The pressure load determined by using formula (1) was introduced for each particular case, as shown in **Figure 20**.

The mesh has been generated as illustrated in **Figure 21**. The selected shape of the finite elements was tetrahedral, with an approximate global size of 0.25 mm and a maximum curvature control having a deviation factor of about 0.1 mm.

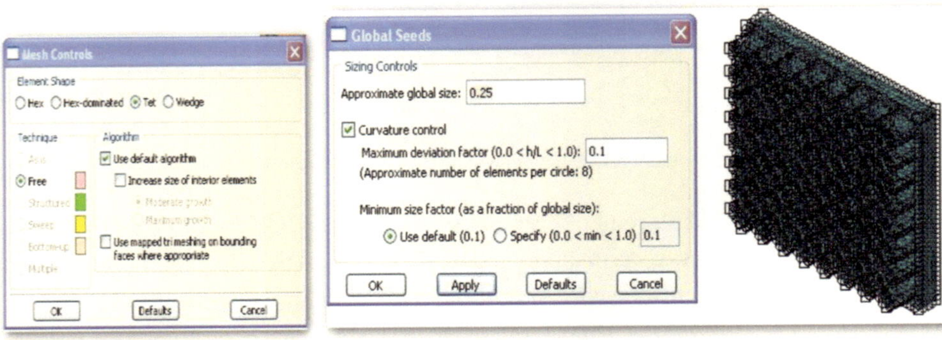

Figure 21. Element-type control and mesh-size control defined for the FEA.

3.4. Results of the finite element analyses made with Abaqus

The equivalent von Mises stress, displacement and equivalent strain were determined for all the six samples presented in **Figure 18**, using the finite element analysis method. The analysis allowed formulating recommendations with reference to the optimum values ranging in the standard limits that exist in this field.

In order to determine the equivalent von Mises stress, for the calculus made within Abaqus FEA program, it has been considered the fifth theory for multiaxial stresses:

$$\sigma_{ech} = \sqrt{\frac{1}{2}[(\sigma_x - \sigma_y)^2 + (\sigma_y - \sigma_z)^2 + (\sigma_z - \sigma_x)^2 + 3(\tau_{xy}{}^2 + \tau_{yz}{}^2 + \tau_{zx}{}^2)]} \geq 0; \sigma_{ech} \leq \sigma_c \quad (2)$$

If we consider in formula (2) that the tangential stresses are equal,

$$\begin{cases} \tau_{xz} = \tau_{zx} \\ \tau_{yz} = \tau_{zy} \\ \tau_{xy} = \tau_{yx} \end{cases} \quad (3)$$

a particular state of uniaxial stress is obtained. The simplified formula was used within Abaqus FEA program in order to perform the interpretation of the stress results.

The stress energy was also calculated according to formula (4):

$$E_{def} = \frac{1}{2}\sigma_{ech} \times \varepsilon_{ech} \tag{4}$$

where E_{def} - represents the stress energy; σ_{ech} - equivalent stress von Mises; ε_{ech} - equivalent strain.

As one may notice in **Figure 22**, the values of the equivalent stress von Mises were lower than the yield strength σ_c= 775 MPa specified in the material datasheet provided by the supplier of the TiAl6V4 material. The minimum value of the equivalent stress von Mises (102.50 MPa) has been determined in the case of the sample with lattice structure designed as having cells with a rhombic shape, while the maximum value (278.56 MPa) was obtained in the case of the sample designed with cells having a conical shape. If we consider a value of 212 MPa for the mechanical resistance of the trabecular bone, the adequate solution would be to design and produce the medical implant as having a rhombic shape of the cell, in this case [25].

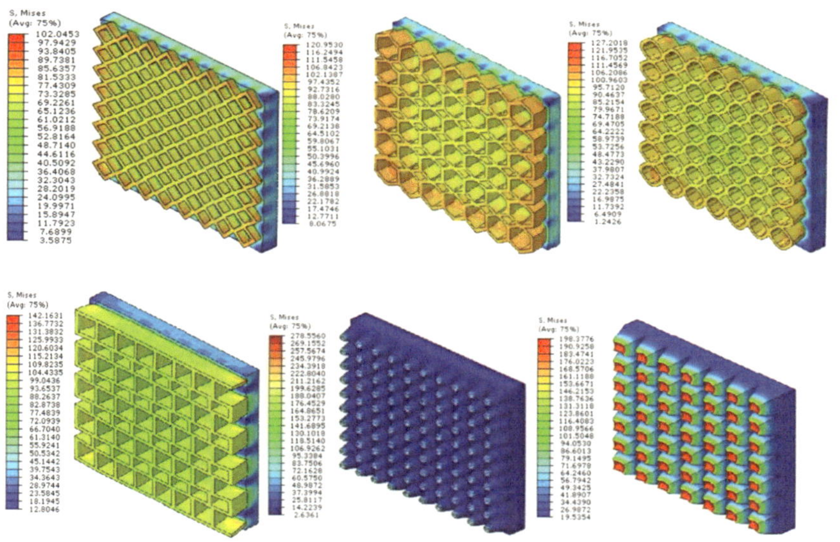

Figure 22. Distribution of the equivalent stress von Mises as computed by the Abaqus FEA program.

The displacement and equivalent strain values were negligible in all cases that were analyzed, being considered too low for affecting the bone or customized medical implant made from titanium powder by SLM [25].

3.5. Manufacturing samples with lattice structures and medical implant by SLM

The manufacturing of the six samples was realized using the MCP Realizer II SLM 250 from the National Centre of Innovative Manufacturing from the Technical University of Cluj-Napoca (TUC-N) presented in **Figure 2**. After the metallic supports were removed, the manufactured samples made by SLM were obtained as presented in **Figure 23**.

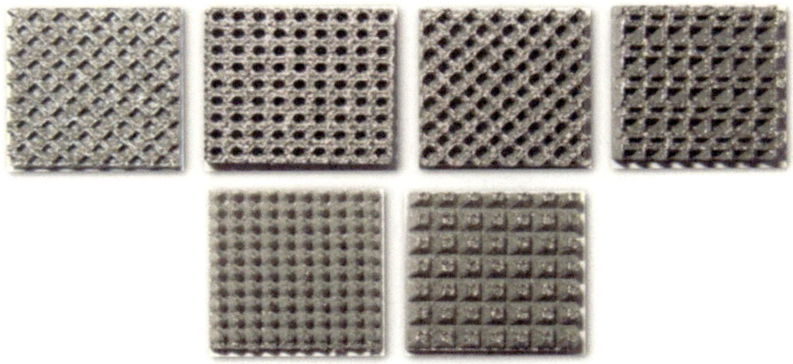

Figure 23. Samples manufactured using the MCP Realizer II SLM 250 equipment [26].

After the manufacturing process, the structure of the samples was analyzed by using a Scanning Electron Microscope (SEM) JSM-5600 LV (JEOL) – type that is available at the Technical University of Cluj-Napoca (TUC-N).

The best geometrical configuration of the lattice structures that were manufactured by SLM was the rhombic one (see **Figure 24**). By analyzing the image presented in **Figure 24.a**, a series of appreciations regarding the metallic grains distribution on the surface of the sample was made. As it is possible to observe in this image, the grains are distributed in a uniform manner.

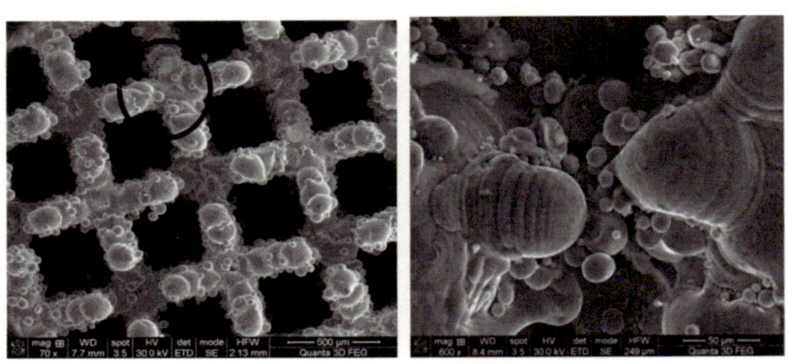

(a) Sample realized by SLM (70 X) (b) Sample realized by SLM (600 x)

Figure 24. Metallographic analysis of the sample with rhombic geometrical shape manufactured by SLM. (a) Sample realized by SLM (70X) (b) Sample realized by SLM (600 x).

By zooming in the surface analyzed area (600x), it was possible to observe that the presence of secondary pores was at the lowest level in this case, conferring higher resistance connections of the grains in the lattice structure. By analyzing the image presented in **Figure 24b** it is possible to observe that the grains that were partially melted formed bridge connections with the grains stated in a semi-liquid phase of about 50 μm.

Taking into account the obtained results, a customized medical implant was manufactured by SLM from TiAl6V4 material for a German Medical Institute, using the SLM 250 HL equipment from SLM Solutions Gmbh Company from Luebeck, Germany, presented in **Figure 8**.

This equipment was selected for the manufacturing process of the customized implant presented in **Figure 25**, due to the fact that this type of equipment, as compared to the MCP Realizer SLM 250 equipment from the Technical University of Cluj-Napoca allows the possibility of using different type of scanning strategies in the same deposited layers, in such way that at the end, in some areas of the implant (e.g. the area where fixing screws are required) this areas will result with a density of 100%, while in other areas (e.g. areas where the medical implant is getting in contact with the human tissue), the areas will be a porous one, the osseointegration process being facilitated in a significant way in that region.

Manufactured area	Scanning strategy	Laser power [W]	Scanning speed [mm/s]	Layer thickness [μm]
Supports	Hatch Solid	100	500	30
Solid areas	Stripes with skin	100	550	30
Porous structure	Stripes with skin	175	470	30

Table 5. Technological parameters used for manufacturing the customized medical implants by SLM using the SLM 250 HL equipment.

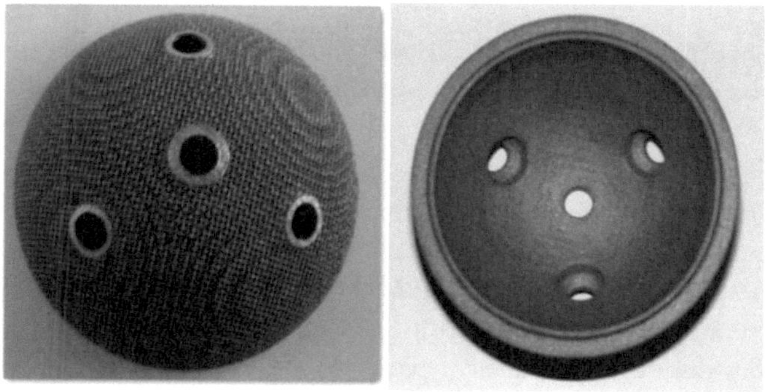

Figure 25. Customized medical implant manufactured from Ti6Al4V material using the SLM 250 HL equipment.

A metallographic analysis of the medical implant was made using the JSM-5600 LV (JEOL) Scanning Electron Microscope (SEM) from the Technical University of Cluj-Napoca. The analyzed area was a mixed area that is close to the contact area of the fixing screws (an area with 100% density of the material) and an area with a controlled porosity given by the geometrical shape of the lattice structure, as it is possible to observe in **Figure 26**.

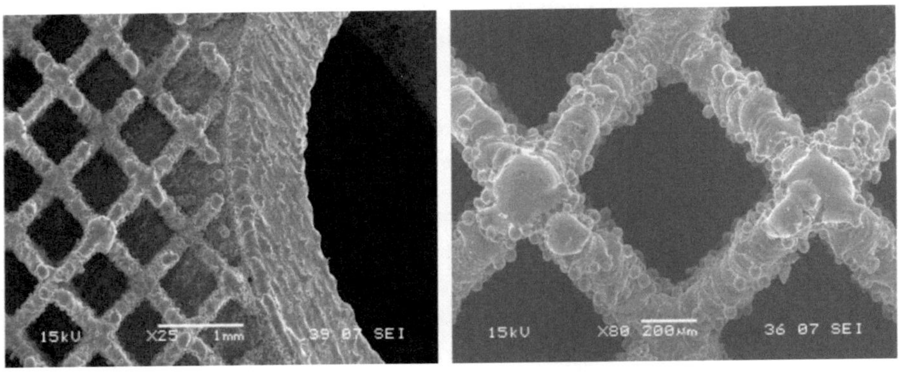

Figure 26. (a) Lattice structure (2x), (b) welding lines (25x) of the customized medical implant made by SLM.

As it is possible to observe in **Figure 26.a**, the laser beam has followed very accurately the geometrical path of the bore that was designed for fixing the implant with screws. The density in this area is 100%, conferring a higher mechanical strength that is required for fixing the medical implant onto the femoral head.

By zooming in the image of the analyzed area, as it is possible to observe in **Figure 26.b**, the size and the distribution of the pores resulted in the structure of the material was homogenous and uniformly distributed (wall thickness of the cells was approximately 1 mm), an aspect that is important for the future proliferation of the human tissue within the structure of the medical implant within the osseointegration process, at the end.

3.6. Conclusions

As it was possible to observe by analyzing the results obtained after the finite element analyses, the sample with the cells having a rhombic shape has proved to be optimal from the mechanical resistance point of view as compared to the results obtained in the case of the other analyzed samples. Taking into account the obtained results, a customized medical implant was manufactured by SLM from TiAl6V4 material for a German Medical Institute, using the SLM 250 HL equipment from SLM Solutions GmbH Company from Luebeck. A metallographic analysis of the medical implant was made using the JSM-5600 LV (JEOL) Scanning Electron Microscope (SEM) from the Technical University of Cluj-Napoca, proving the fact that the laser beam has followed very accurately the geometrical path of the bore that was designed for fixing the implant with screws, and that the size and the distribution of the pores resulted in the structure of the material which is homogenous and uniform, with positive consequences for the future proliferation of the human tissue within the structure of the medical implant within the osseointegration process. Further researches are required to be done in the future regarding the possibilities of manufacturing customized medical implants with a well-controlled level of porosity, made from new types of biocompatible materials (e.g. titanium samples coated with hydroxyapatite or PMMA). Finding a proper method for the stress-release of customized medical implants made from titanium based alloys, during or after the Selective Laser Melting (SLM) process still presents an important challenge for the future researches, as well.

Author details

Pacurar Razvan* and Pacurar Ancuta

*Address all correspondence to: razvan.pacurar@tcm.utcluj.ro, ancuta.costea@tcm.utcluj.ro

Faculty of Machine Building, Department of Manufacturing Engineering, The Technical University of Cluj-Napoca, Cluj-Napoca, Romania

References

[1] Păcurar, R., Păcurar, A., "finite element analysis to improve the accuracy of parts made by stainless steel 316L material using selective laser melting technology", Applied Mechanics and Materials, vol. 657, (2014), pp. 236–240

[2] Tolosa, I., Garciandía, F., Zubiri, F., Zapirain, F., "Study of mechanical properties of AISI 316 stainless steel processed by "selective laser melting", following different manufacturing strategies", International Journal of Advanced Manufacturing Technology, vol. 51, (2010), pp. 639–647

[3] Badrossamay, M., Yasa, E., Van Vaerenbergh, J., Kruth, J.P., "Improving productivity rate in SLM of commercial steel powders", Proceedings of SME RAPID 2009 Schaumburg, Illinois, USA, 12-14 May, 2009

[4] Mahesh, M., Wong, Y., "Benchmarking for Comparative Evaluation of RP Systems and Processes", Rapid Prototyping Journal, vol. 10, Issue 2, (2004), pp. 123–135

[5] Childs, T.H.C., Hauser, C., and Badrossamay, M., "Mapping and modelling single scan track formation in direct metal selective laser melting", CIRP Annals – Manufacturing Technology, vol. 53 (1), (2004), pp. 191–194

[6] Dong-Gyu A., "Applications of laser assisted metal rapid tooling process to manufacture of moulding & forming tools — state of the art", International Journal of Precision Engineering and Manufacturing, vol. 12 (5), (2011), pp. 925–938

[7] Morovič, L., Pokorný, P., "Optical 3D Scanning of Small Parts", Advanced, Materials Research, vols. 468-471, (2012), pp. 2269–2273, Trans Tech Publications, Switzerland, doi: 10.4028/www.scientific.net/AMR.468-471.2269

[8] Gower, H.L., Pieters, R.R.G.M., Richardson, I.M., "Pulsed laser welding of metal–polymer sandwich materials using pulse shaping", Journal of Laser Applications, vol. 18 (1), (2005), pp. 35–41

[9] Kruth, J.P., Froyen, L., Van Vaerenbergh, J., Mercelis, P., Rombouts, M., Lauwers, B., "Selective laser melting of iron-based powder", Journal of Materials Processing Technology, vol. 149 (1–3), (2004), pp. 616–622

[10] Kruth, J.P., Levy, G., Klocke, F., Childs, T.H.C., "Consolidation phenomena in laser and powder-bed based layered manufacturing", CIRP Annals-Manufacturing Technology, vol. 56 (2), (2007), pp. 730–759

[11] Levy, G.N., Schindel, R., Kruth, J.P, "Rapid manufacturing and rapid tooling with layer manufacturing (LM) technologies, state of the art and future perspectives", Annals of the CIRP, vol. 52 (2), (2003), pp. 589–609

[12] Li, R., Shi, Y., Liu, J., Yao, H., Zhang, W., "Effects of processing parameters on the temperature field of selective laser melting metal powder", Powder Metallurgy and Metal Ceramics, vol. 48 (3–4), (2009), pp. 186–195

[13] Moures, F., Cicala, E., Sallamand, P., Grevey, D., Vannes, B., Ignat, S., "Optimization of refractory coatings realised with cored wire addition using a high-power diode laser", Surface and Coatings Technology, vol. 200 (7), (2005), pp. 2283–2292

[14] Sparks, T., Ruan, J., Fan, Z., Bap, Y., Liou, F., "Effect of structured laser pulses on grain growth in H13 tool steel", Solid Freeform Fabrication Symposium, Austin, TX, (2006), pp. 261–267

[15] Tolochko, N.K., Mozzharov, S., Yadroitsev, I.A., Laoui, T., Froyen, L., Titov, V.I., Ignatiev, M.B., "Balling process during selective laser treatment of powders", Rapid Prototyping Journal, vol. 10 (2), (2004), pp. 78–87

[16] AISI Type H13 Hot Work Tool Steel material characteristics http://www.matweb.com/search/DataSheet.aspx?MatGUID=e30d1d1038164808a85cf7ba6aa87ef7&ckck=1, accessed: 15.09.2015

[17] SLM Solutions GmbH, SLM Materials Characteristics, available at: http://www.slm-solutions.com/cms/upload/pdf/120923_SLM_Materialien.pdf, accessed: 15.09.2015

[18] Păcurar, R., Păcurar, A., Petrilak A., Bâlc. N., "Finite element analysis to predict the mechanical behaviour of lattice structures made by selective laser melting technology", Applied Mechanics and Materials, vol. 657, (2014), pp. 231–235

[19] Lin, C.Y., Kikuchi, N., Hollister, S.J., "A novel method for biomaterial scaffold internal architecture design to match bone elastic properties with desired porosity", Journal of Biomechanics vol. 37 (5), (2004), pp. 623–636

[20] Maričić, S., Pavičić, D.K., Perinić, M., Lajnert, V., "The use of technological documentation in vestibuloplasy fixture plate production", Medicina Fluminensis, vol. 47 (3), (2011), pp. 294-298

[21] Marcincinova, L., N., Fecova, V., Marcincin, J.N., Janak, M., Barna, J., "Effective utilization of rapid prototyping technology", AIP Conference Proceedings, vol. 1431, (2012), pp. 834-841

[22] Marcincinova, L.N, Marcincin, J. N., "Selected testing for Rapid Prototyping technology operation", Applied Mechanics and Materials, vol. 308, (2013), pp. 25-31

[23] Yan, C., Hao, L., Hussein, A., Raymont, D.,"Evaluations of cellular lattice structures manufactured using SLM", International Journal of Machine Tools & Manufacture, vol. 62, (2012), pp. 32–38

[24] Hussein, A., Hao, L., Yan, C., Everson, R., Young, P.,"Advanced lattice support structures for metal additive manufacturing", Journal of Materials Processing Technology, vol. 213, (2013), pp. 1019–1026

[25] Toan, T. L., MD, "Biomechanics of Fractures and Fixation", 2005, available at: http://www.docslide.com/biomechanics-of-fractures-and-fixation/, accessed: 18.09.2015

[26] Mager, V., "Research regarding the manufacturing of customized medical implants made from composite materials, by using the selective laser melting technology" PhD Thesis, Technical University of Cluj-Napoca, 2014

Chapter 9

On the Role of Interfacial Reactions, Dissolution and Secondary Precipitation During the Laser Additive Manufacturing of Metal Matrix Composites: A Review

Anne I. Mertens and Jacqueline Lecomte-Beckers

Additional information is available at the end of the chapter

http://dx.doi.org/10.5772/63045

Abstract

Since current trends in the transportation, energy or mechanical industries impose increasingly demanding service conditions for metallic parts, metal matrix composites (MMCs) are the object of a growing interest. Powder-based laser additive manufacturing, which allows making parts with complex shapes, appears particularly adapted for the production of MMCs. This paper reviews the current state-of-the-art in the production of MMCs by additive processes, with the aim of assessing the potentials and difficulties offered by these techniques. Two main processing routes are envisaged, i.e. (1) the processing of ex situ composites in which the reinforcing phase as a powder—often of ceramic particles—is directly mixed with the powder of the matrix alloy, and both powders are simultaneously processed by the laser. (2) Alternatively, the reinforcing phase can be produced in situ by a chemical reaction during the fabrication of the composite. For both processing routes, a careful control is needed to overcome challenges brought, e.g. by the behaviour of the reinforcement particles in the laser beam, by changes in laser absorptivity or by the dissolution of the reinforcing particles in the molten metal, in order to produce MMCs with enhanced usage properties.

Keywords: metal matrix composites, interfacial reactions, solidification, additive manufacturing, laser processing

1. Introduction

Current requirements for improved efficiency in the mechanical and energy industries impose ever more demanding service conditions for metallic parts, e.g. higher service temperatures, corrosive environments and/or increased mechanical loads. At the same time, demands for metallic materials with enhanced specific properties are also increasing in view of the trends for lightweighting in portable applications. In this context, metal matrix composites (MMCs), combining the advantages of the metallic matrix with the beneficial contribution of a well-selected second phase, appear as materials of choice that can be designed and tailor-made in view of a specific application, in bulk or as coating [1].

The vast choice of potential second phases opens unlimited possibilities in terms of the usage properties that can be attained. Indeed, the reinforcements may take on different morphology (i.e. long fibres, short fibres or particles) and size (i.e. in the micro- or nano-size range) [1]. Various reinforcements may even be combined to make a hybrid composite [1, 2]. Among the most popular types of reinforcements, carbides such as tungsten (WC) [3–7], chromium (Cr_3C_2) [3, 8], silicon (SiC) [2, 9–12] or titanium carbides (TiC) [13–15] have often been used in view of their high hardness to enhance the wear resistance of the composites. Oxides [16, 17], nitrides [18] or borides [19, 20] also proved of interest as reinforcement, as did intermetallics [9, 15]. Alternatively, the second phase may also be selected in order to fulfil a specific function, such as, self-cleaning [21], self-healing [22, 23] or as solid lubricant in self-lubricating MMCs that are currently attracting a growing interest for applications where classical lubrication methods do not work [23–26].

A number of different methods can be used for the fabrication of MMCs. Melting metallurgical processes include infiltration of a preform by squeeze casting [1, 27], reaction infiltration or stir casting [1]. Powder metallurgy processes involve sintering, pressing or forging of a mixture of powders or of composite powders [1, 24], while severe plastic deformation processes such as friction stir processing rely solely on solid-state material flow [27, 28]. These conventional processes for the elaboration of MMCs share a common limitation. Indeed, it is very difficult to fabricate MMC components with complex shapes by these methods [2]. On the other hand, additive manufacturing and particularly powder-based additive techniques offer the possibility to fabricate any complex geometry directly from the powders [2, 29]. General features of additive manufacturing processes suitable for the fabrication of MMCs will be reviewed in more details in the second section of this chapter, while Sections 3 and 4 will focus on some specific examples of MMCs processed by additive manufacturing, along with their properties and envisioned applications.

2. Additive manufacturing techniques for the production of metal matrix composites

Following standard ASTM F2792-10, additive manufacturing is defined as *"the processing of joining materials to make objects from 3D model data, usually layer upon layer, as opposed to subtractive manufacturing technologies"*. This definition encompasses a large variety of manufacturing processes suitable for the fabrication of MMCs (**Table 1**). 3D printing that relies on a binding agent to consolidate a powder has been used successfully to produce polymer matrix composites with particulate reinforcements [29]. This technique can also be applied to the manufacturing of metallic materials, but due to the intrinsic nature of its consolidation process [29, 30], it does not allow for making as dense products as can be obtained from polymers. In this case, the advantage of additive manufacturing lies in the excellent dimensional accuracy of the part [30] and in the possibility to design and fabricate directional 3D preforms with optimised out-of-plane mechanical properties [31]. Fully dense parts can only be obtained by further infiltrating the porous preform [29, 30]. This process is thus very similar to the conventional processing of MMCs by infiltration or squeeze casting, and it will not be discussed in more details in this chapter.

Classification	Technology	Description	Raw materials	Comment
Sheet lamination	Laminate object manufacturing (LOM) Ultrasonic consolidation (UC)	Builds part by joining sheets or foils of materials	Laminates or foils	Composite laminates or sheets must be prepared prior to the fabrication of a 3D part.
Binder jetting	3D printing	Builds parts by depositing locally a binding agent	Powders	This process is suitable for the production of porous preforms. Further infiltration is needed to make fully dense composites.
Beam deposition	Direct metal deposition (DMD) Laser deposition Laser cladding	Fusing a material with a laser during simultaneous deposition on a substrate	Powders or wires	Suitable for the fabrication of MMCs including functionally graded materials (FGMs).
Powder-bed Fusion	Direct metal laser sintering (DMLS) Selective laser melting (SLM) Laser beam melting (LBM) Selective laser sintering (SLS)	Uses a laser to melt a powder bed at desired locations	Powders	Suitable for the fabrication of MMCs with a constant volume fraction of second phase.

Table 1. Classification of additive manufacturing processes for the production of metal matrix composites (MMCs) [12, 29, 32].

Laminated object manufacturing allows the fabrication of MMCs reinforced with either particulates or fibres (**Table 1**) [29]. However, this process requires the production of composite sheets prior to making the composite 3D part, as is also the case for ultrasonic consolidation in which metallic—or MMC—foils are joined together by ultrasonic welding (**Table 1**) [29]. Both these methods offer the possibility to control precisely the placement of the reinforcement inside the final component, but they are not particularly well adapted to the fabrication of MMCs parts with complex geometries.

When it comes to making MMCs components with complicated shapes, powder-based additive manufacturing techniques demonstrate the greatest potential, in spite of being generally limited to particulate reinforcements (**Table 1**) [12, 29]. Powder-based additive manufacturing techniques can be classified into two broad categories, i.e. (1) beam deposition processes in which a high-energy beam, typically a laser, is used to fuse a powdered material as it is projected onto a substrate and (2) powder-bed fusion processes that use a focused energy beam to melt locally selected zones of a powder bed. The first category encompasses the processes known as direct metal deposition (DMD), laser deposition (LD), laser engineering net shaping (LENS) or laser cladding (LC)—the latter denomination can also be used to designate the laser surface treatment of pre-deposited powdered layer. Direct metal laser sintering (DMLS), laser beam melting (LBM) and selective laser melting (SLM) are examples of the second category [12, 29, 32].

From a practical viewpoint, powder-bed fusion processes require that the two constituents of the composite, i.e. the matrix and reinforcing powders, should be pre-mixed prior to their spreading in the powder bed [2]. In beam deposition processes, on the other hand, it is also possible to feed the two powders into the beam from two separate hoppers, without pre-mixing. This latter set-up is favourable in order to avoid any segregation of the constituents due to difference in the densities of the powders that could impair a precise compositional control of the composite material [15]. Moreover, it opens the possibility of gradually varying the ratio of the matrix and reinforcement powders during processing so as to make functionally graded materials (FGMs) [10, 11, 15, 18]. LC of pre-deposited pre-mixed powders may still be preferred in some occasions, e.g. for depositing a composite layer on a substrate characterised by a very high reflectivity [26]. Finally, it is worth noting that while the vast majority of the research on the production of MMCs by beam deposition processes has been focused on powder-feed techniques, few investigations have also been carried out into the fabrication of composites by a mixed wire- and powder-feed process [6, 33].

Practical differences set aside, all powder-based laser additive manufacturing processes are faced with similar issues and challenges during the fabrication of MMCs reinforced by ceramic particles. These challenges are related

1. to interactions between the ceramic particles and the laser beam: In a few instances, the reinforcing particles have been reported to melt or decompose when exposed to the action of the laser [25, 34]. Besides, absorptivity of the laser beam by ceramic particles may differ significantly from the absorptivity by metallic powders, thus affecting the transfer of energy from the laser to the built [2, 5, 14, 15, 35]. As a consequence, it may prove necessary to adjust continuously the processing parameters as a function of the volume fraction of ceramic particles during the fabrication of FGMs [15]. Alternatively, Przybylowicz and Kusinski [5] suggested to coat WC or TiC particles with nickel to lower down their absorption coefficient to values comparable to the absorption coefficient of the metallic powders.

2. to interactions between the ceramic particles and the metallic melt pool: while a few examples of matrix-reinforcement pairs exist for which the dissolution of the reinforcing particles in the melt pool remains very limited [19], most often ceramic particles tend to dissolve at least partially in the melt pool [8, 11–13], as illustrated in **Figure 1** in the case of a stainless steel 316L-WC composite [36]. Consequently, new phases may form as a result of secondary precipitation inside the metallic matrix. Due to the ultra-fast thermal cycles typically imposed by laser additive manufacturing processes, these new phases generally exhibit ultra-fine structures often resulting in enhanced hardness and wear properties [2, 11, 35, 36]. Interfacial reactions, dissolution and secondary precipitation phenomena have even been used to synthesise the reinforcing phase of MMCs by in situ reactions. In this case, the energy of the laser is not only used to fuse the powdered materials but also to overcome the activation energy barrier for the reaction and form new chemical compounds [2, 29, 37–41].

Figure 1. SEM micrograph of a stainless steel 316L composite coating with 16 vol% of tungsten carbides deposited by laser cladding [36].

Either way, in both the ex situ and the in situ processing of MMCs, a careful control of the reactions taking place during laser additive manufacturing is needed in order to produce MMCs with enhanced properties. The following sections of this chapter thus aim at reviewing systematically pairs of metallic matrix and reinforcing phase that have already been produced by additive manufacturing, challenges that were met during the production of these composites and the properties that could be attained. Section 3 focuses on the fabrication of ex situ MMCs, and Section 4 on the in situ synthesis of MMCs. A synoptic table summarising the major characteristics of these MMCs is proposed in Section 5. Special care is also taken to identify current fundamental issues that should more particularly be the object of future work.

3. Ex situ MMCs produced by laser additive manufacturing

3.1. Aluminium matrix composites

3.1.1. Aluminium matrix composites reinforced with carbides

A major concern during the fabrication of aluminium (Al) matrix composites reinforced with carbide particles is to avoid the formation of aluminium carbide Al_4C_3 through interfacial reactions between the liquid Al and the particles [6, 13, 42]. Indeed, the presence of Al_4C_3 in the composite is potentially detrimental to both its mechanical and its corrosion resistance, as it is well known that Al_4C_3 is extremely brittle [13] and susceptible to hydrolysis in humid environment, thus giving rise to high corrosion rate [6]. A number of different strategies have been devised with various levels of success in order to solve this problem, including the careful selection of the carbide particles used as reinforcement [2, 13], the chemical modification of the matrix alloy [6, 12] or the optimisation of the processing conditions, e.g. to promote the formation of less deleterious aluminium carbides such as the mixed aluminium-silicon carbide Al_4SiC_4 [2, 19, 43, 44].

In their investigation into the fabrication by DMLS of AlSi10Mg matrix composites with 10 wt % of SiC particles, Manfredi et al. [12] relied on the earlier observation that the addition of a high amount (i.e. of at least 7 at% [42]) of silicon in the matrix alloy could suppress the reaction between the Al alloy and the SiC particles leading to the formation of Al_4C_3 during the fabrication of Al matrix composites by conventional processes such as melt infiltration. The experimental parameters for the DMLS process were set as follows: the scan speed was varied between 500 and 700 mm/s, the laser power between 180 and 195 W, the layer thickness was set at 30 μm and the hatching distance at 0.17 mm. Although the hardness of the composite materials was significantly increased in comparison with AlSi10Mg processed in the same conditions, analysis of the AlSi10Mg/SiC composites by X-Ray Diffraction (XRD) showed that the SiC particles had almost fully disappeared while a significant amount of Al_4C_3 had been formed [12]. The behaviour of the DMLS AlSi10Mg/SiC composites was thus found to differ significantly from their conventional counterparts.

Also investigating the fabrication of AlSi10Mg/SiC composites by SLM, Chang et al. [2] used the same layer thickness of 30 μm as used by Manfredi et al. [12], with a smaller hatch spacing of 0.05 mm, a lower laser power of 100 W and a lower scan speed of 100 mm/s. Dissolution of the SiC particles was also observed in their study, but this time, the reaction product was the mixed Al_4SiC_4 carbide, that has the advantages of being chemically inert in humid environment and of being less brittle than Al_4C_3. By considering the energy density per unit volume for both studies, it is found that the set of experimental parameters used by Manfredi et al. [12] resulted in a much lower value of the energy density of 77 J/mm^3 when compared to the set of processing parameters of Chang et al. [2] giving rise to an energy density of 666 J/mm^3. As a consequence, the samples of Chang et al. [2] experienced a greater heat accumulation during fabrication and a higher temperature when compared with the samples of Manfredi et al. [12]. The processing strategy of Chang et al. [2] thus proved successful in favouring the formation of Al_4SiC_4 that is expected to form at temperatures above 1350–1400°C, whereas the formation of the deleterious Al_4C_3 cannot be avoided at lower processing temperatures [2, 19], in spite of the high silicon content of the matrix alloy [12].

Ghosh et al. [43, 44], on the other hand, investigated the fabrication of Al-4.5Cu-3Mg/SiC composites by DMLS using a pulsed Nd:YAG laser while varying the SiC volume fraction and particles size. However, due to the strongly non-uniform nature of the temperature distribution created by the pulsed laser, their composites were subjected to high internal stresses that resulted in extensive cracking in all their specimens [44, 45].

In a similar way to silicon carbides, tungsten carbides (WC) present a strong tendency to dissolve in Al alloys during LD, leading to the formation of Al–W intermetallics and of the detrimental Al_4C_3 [6]. In order to avoid the formation of the latter phase, Li et al. [6] attempted to modify the chemical composition of their Al-11 wt% Si matrix alloy by adding pure elemental Ti powder together with the WC particles. The addition of Ti in the melt pool was found successful in inhibiting the formation of the deleterious Al_4C_3 by promoting the precipitation of titanium carbides (TiC) both as dispersed phase within the Al matrix and as a passivation layer at the interface between the WC particles and the Al matrix. Mixed Al–Ti (Al_3Ti or AlTi) and Al–W ($Al_{12}W$ or Al_4W) intermetallic compounds were also detected by XRD.

On the other hand, titanium carbides (TiC) present a much lower reactivity with Al than SiC and WC [13]. Indeed, TiC possesses a low solubility in Al, and its dissolution rate is also slower than for SiC. Coatings with a matrix of Al-12 wt% Si and 40 wt% TiC as reinforcement were successfully produced by LC [13]. Very limited dissolution of the TiC particles into the Al matrix led to the formation of a small amount (about 3%) of finely dispersed Ti_3SiC_2, whereas the formation of the deleterious Al_4C_3 was completely avoided. Both the hardness and the sliding wear resistance of the composite coating were significantly improved by the TiC addition.

In view of their excellent thermodynamic stability, TiC nanoparticles have also been used as reinforcement in AlSi10Mg matrix composites [46–49]. Due to their strong tendency to form clusters and agglomerates [22, 29], nanoparticles may reflect poorly on the flowability and/or the spreadability of the powders, and nanostructured or nanocomposite powders of micron-size are usually preferred for the additive manufacturing of nanocomposites. In various studies [46–49], careful mixing of TiC nanoparticles with an average size of 50 nm and of AlSi10Mg particles with an average size of 30 μm in a planetary ball mill resulted in the uniform distribution of the TiC nanoparticles around the surface of the AlSi10Mg particles, thus forming nanostructured particles suitable for further processing by SLM.

In a first set of studies [46, 47], the effect of the linear laser energy density on parts density, microstructure, hardness and wear behaviour was investigated by adjusting the scan speed at a constant laser power of 110 W. The second phases present in the composites were identified as TiC and Mg_2Si, the latter formed by natural aging of the AlSi10Mg matrix alloy. Moreover, the highest investigated linear laser energy density of 733 J/m was found to result in parts with an optimum relative density of 98%, uniform distribution of the TiC nanoparticles throughout the composite and optimised hardness and wear resistance. Dai and Gu [48] reported on the optimisation of the surface finish of AlSi10Mg/TiC nanocomposites by varying the scan speed at a constant laser power of 150 W. In this case, medium values of the linear laser energy density gave the best result. Lower linear laser energy densities resulted in an excessive viscosity and poor spreading of the melt pool on the underlying solid surface [46, 47]. Higher linear laser energy density, on the other hand, led to excessive vaporisation of the melt [48]. Both situations resulted in a smaller melt pool and in a poorer surface finish. Finally, in an attempt to optimise the tensile properties of TiC/AlSi10Mg nanocomposites produced by SLM, Gu et al. [49] reported on an investigation of the influence of the processing parameters on the dispersion of the nanoparticles. Gu et al. [49] demonstrated the feasibility of tailoring the spatial distribution of the TiC nanoparticles in the composite by means of a careful control of the balance of the torque forces in the melt pool and of repulsive capillary forces arising between the nanoparticles under specific processing conditions.

3.1.2. Aluminium matrix composites reinforced with borides

In opposition to carbides and particularly to SiC or WC whose reactions with liquid Al may result in the formation of brittle and deleterious compounds, titanium diboride (TiB_2) is attracting a growing interest for use as reinforcement in Al matrix composites due to its very low reactivity with Al. Anandkumar et al. [19] thus investigated the LC of an Al-12 wt% Si alloy with 40 wt% TiB_2. Using processing parameters similar to those reported for the LC of Al-12 wt% Si/TiC composites [13], the Al-12 wt% Si/TiB_2 composites did not present any sign of dissolution of the TiB_2 particles, nor of interfacial reactions between the particles and the Al matrix [19]. Moreover, when compared with the matrix alloy, the wear resistance of the Al-12 wt% Si/TiB_2 composite coating, as characterised by dry sliding wear tests using a counterbody of quenched and tempered AISI440C tool steel, was greatly increased by the addition of the TiB_2 particles. Being much harder than the counterbody (2550 HV vs 800 HV), protruding TiB_2 particles proved very efficient in supporting the contact stresses with the counterbody and protecting the Al-12 wt% Si matrix from continued intense plastic deformation.

3.1.3. Aluminium matrix composites reinforced with oxides

Little information is available in literature on the laser additive manufacturing of Al matrix composites with ex situ oxide particles. Manfredi et al. [12] investigated the fabrication of AlSi10Mg matrix composites with nanoparticles of the aluminium-magnesium spinel $MgAl_2O_4$ using DMLS, with limited success. In order to avoid processability issues caused by the tendency of nanoparticles for clustering [22, 29], a nanostructured powder was prepared by mixing the AlSi10Mg and the $MgAl_2O_4$ powders using a ball milling system for 48 hours prior to conducting the DMLS experiments. Even the best of the produced composite specimens still exhibited a non-negligible volume fraction of residual porosity ranging from 2.2 to 3.5%. The microstructure of the composites was also found to be much more inhomogeneous than for the AlSi10Mg alloy processed under the same conditions without spinel nanoparticles, thus suggesting that the spinel nanoparticles might affect the solidification process during

DMLS. And finally, the hardness of the AlSi10Mg/spinel nanoparticles composite was decreased by 11% in comparison with the AlSi10Mg DMLS samples without nanoparticles.

3.2. Ferrous matrix composites

3.2.1. Ferrous matrix composites reinforced with carbides

Silicon carbides have been extensively used as reinforcement in mild [10], medium-carbon [50] or stainless steels [11, 51–53]. A compositionally graded mild steel/SiC composite coating produced by LC was found to exhibit enhanced hardness, wear and corrosion resistance when compared to the bare mild steel substrate [10]. The improvement of the mechanical properties was ascribed not only to the addition of the SiC particles but also to the secondary precipitation of iron silicide Fe_2Si consequent to the partial dissolution of the SiC particles. Depending on the processing conditions, Fe_3Si or Fe_3Si_5 have also been observed [50, 51], and, when SiC particles were dispersed in stainless steel, chromium carbides such as Cr_3C_2 [11, 51, 53] or Cr_7C_3 [52] were also formed. This is a major issue for the laser processing of stainless steel/carbides composites, since an excessive precipitation of chromium carbides might deplete the chromium content of the stainless steel matrix hence compromising its corrosion resistance [3]. However, provided the precipitation of chromium carbides remains limited, Dutta Majumdar and Li [53] have demonstrated that stainless steel 316L/SiC composites could achieve a corrosion resistance that is equivalent or better when compared with conventional stainless steel 316L, in combination with improved hardness and wear resistance [11, 53].

Tunsgten carbides (WC) have also been widely used as reinforcement in steels, and like SiC, they exhibit a strong tendency to dissolve in the ferrous matrix [3, 35, 51, 54], giving way to the secondary precipitation of finely dispersed mixed carbide phases as, e.g. M_6C, $M_{23}C_6$ or M_7C_3 where M stands for Fe, W or Cr (in stainless steel matrix composites). WC particles were found extremely efficient in increasing the hardness and wear resistance of steel matrix composites due to the combined hardening effect of the surviving particles, of the finely dispersed secondary precipitates and of the solid solution strengthening of W dissolved in the steel matrix [3, 35, 51]. Since it is known that the uniform distribution of very fine hard phases may be beneficial in view of improving the cavitation erosion resistance of stainless steel, Lo et al. [35] added deliberately small (~1μm) WC particles in a stainless steel 316 composite coating, so as to favour the complete dissolution of the WC particles and the reprecipitation of finely dispersed carbides. However, as mentioned earlier, secondary precipitation of mixed carbides may prove detrimental to the corrosion resistance of stainless steel/WC composites if the depletion of chromium in the stainless steel matrix reach such an extent as to reduce the capacity of the material to withstand corrosion, as observed by Betts et al. [3] in the case of a stainless steel/WC coating.

Titanium carbides (TiC), on the contrary, are characterised by a great chemical stability in presence of ferrous melts. Cheng et al. [51], for example, did not observe any significant dissolution or reaction during the LC of a stainless steel/TiC composite coating. However, decreasing the size of the TiC particles could result in their partial dissolution [55]. Due to the difference in density between TiC and stainless steel, the TiC particles tend to float at the surface of the stainless steel melt pool. As a consequence, special care must be taken when injecting the TiC particles in the melt, for fear that the TiC particles may distribute unevenly in the composite [51, 55]. When added to a Fe-36% Ni Invar36 alloy, on the other hand, TiC particles dissolved at least partially [56]. Enrichment of the Invar36 matrix with Ti resulted in the stabilisation of the body-centred cubic (BCC) α structure at the expense of the face-centred cubic (FCC) γ phase, significant increase of the coefficient of thermal expansion of the matrix alloy and the loss of its "invar" property whereas the wear resistance of the composite exhibited very little improvement.

Chromium carbides, finally, tend to dissolve at least partially in ferrous matrix, giving rise to the secondary precipitation of small amount of mixed M_7C_3 or $M_{23}C_6$ carbides [3, 51]. Cr_3C_2 particles did not appear as efficient as WC particles in view of enhancing the hardness and wear resistance of stainless steel matrix composites, but they also proved less deleterious to the corrosion resistance [3]. Although secondary precipitation in stainless steel/Cr_3C_2 composite might still have caused some depletion of Cr in the stainless steel matrix, this phenomenon was less severe than in stainless steel/WC composites.

3.2.2. Ferrous matrix composites reinforced with oxides

Attempts to reinforce ferrous matrix composites with oxide particles have met little success so far. In their investigation into the deposition of a stainless steel AISI304/Al_2O_3 composite coating by LC, Xu et al. [34] found that the Al_2O_3 particles were melted upon heating in the laser beam. Al atoms liberated by the dissociation of Al_2O_3 then dissolved in the stainless steel matrix, with positive effect on the hardness of the coating due to the combined effect of the solid solution strengthening of the austenitic stainless steel matrix by Al and of a partial transformation of the austenite into hard martensite. Betts et al. [3] also reported an extensive dissociation of Al_2O_3 particles during the direct LD of AISI316/Al_2O_3 composite coating. Dissociation of Al_2O_3 was complete at higher laser power, whereas the lower laser power resulted only in a low survival rate of the Al_2O_3. The hardness of the AISI316/Al_2O_3 composite was actually decreased in comparison with the AISI316 base material. The LC of Cr_2O_3 in an austenitic stainless steel matrix [51] also proved unsatisfactory: the composite coating exhibited extensive cracking and a poor bonding with the stainless steel substrate.

3.2.3. Ferrous matrix composites reinforced with other types of particles

The LC of chromium borides CrB_2 in austenitic stainless steel UNS S31603 resulted in a significant increase of the composite layer hardness and of its cavitation erosion resistance, in comparison with the base material [51]. Improved properties were linked not only to the presence of surviving CrB_2 particles but also to the secondary precipitation of finely dispersed chromium borides (CrB and Cr_2B) and iron borides (Fe_2B) following the partial dissolution of the initial CrB_2 particles. Li et al. [57] added CeO_2 nanopowder in laser-cladded FeCrBSi/NbC composites: CeO_2 nanoparticles acted as heterogeneous nuclei, favouring higher solidification rates and enhancing the in situ synthesis of finely dispersed NbC (see also Section 4.2).

Hydroxyapatite (HA), in micro- or nano-size, has been used as addition in stainless steel matrix composites produced by SLM for biomedical applications [58, 59]. Stainless steel 316L presents good mechanical and chemical properties for use in load-bearing implants, but it does not promote bone adhesion or tissue regrowth on the implant. HA, on the other hand, lacks in ductility and toughness for load-bearing applications, but it is very efficient in favouring bone and tissue regrowth. By combining the properties of the two materials, load-bearing and bioactive implants have been fabricated. Since the density of HA is significantly lower than the density of SS 316L, the ceramic particles tend to float at the surface of the melt pool and to aggregate, resulting in insufficient wetting or "balling", and in the increased formation of porosities and cracks in the SS316L/HA composites. Consequently, a careful optimisation of the SLM processing parameters was necessary in order to avoid these problems. Optimised SLM SS316L/nano-HA composites were found to exhibit a higher tensile strength than their bulk SS316L counterparts, in combination with a good ductility [59].

3.3. Nickel and nickel alloys matrix composites

Carbides, and particularly WC, are by far the most widely used ceramic reinforcements in Ni-based composites. WC actually presents a lower reactivity with Ni-based alloys than with ferrous alloys [54]. As a consequence, nickel is sometimes used to coat ceramic powder particles prior to their use as reinforcement, to slow down the dissolution of the ceramic particles in, e.g. ferrous matrix composites [3, 35, 60] and to lower the laser absorptivity of the ceramic to values similar to the laser absorptivity of metallic materials [5]. Due to the low reactivity of WC with Ni-based alloys, it is fairly easy to control the rate of dissolution of the WC particles in Ni-matrix composites by optimising the processing parameters (e.g. by decreasing the laser power) [5, 61]. Partial dissolution of WC in Ni alloys has been reported, leading to secondary precipitation of various types of tungsten carbides (secondary WC, W_2C...) as well as of mixed carbides (M_7C_3, $M_{23}C_6$...) [5, 60, 62]. Ni–W intermetallics (Ni_2W_4C, $Ni_{17}W_3$...) were also observed in NiCrBSi/WC composites processed by LC at high laser power and on a substrate pre-heated up to 400°C [63]. MMCs reinforced with WC have been successfully produced in a wide range of Ni-based alloys: pure Ni [60, 64], NiCrSiFe [65], NiCrBSi alloys [62, 63], Ni35 (containing Cr, Si, B and Fe) [61] or Ni60 (containing also Co) [66].

Addition of chromium carbide Cr_3C_2 in alloy Inconel 625 gave good results in terms of wear resistance, due to the partial dissolution of Cr_3C_2 in the metallic matrix and reprecipitation of fine Cr-rich M_7C_3 carbides [8]. Titanium carbides (TiC) were also deemed very promising for use as reinforcement in Inconel 625 [67] and Inconel 690 [68]. Minimal dissolution of TiC was observed in Inconel 690 [68], leading to the formation of fine dendritic $M_{23}C_6$.

Other reinforcements suitable for use in Ni-based matrix composites include titanium diboride (TiB_2) [26], and rare earth oxides like CeO_2 [17] or Y_2O_3 nanoparticles [69]. Both types of oxide nanoparticles were found very efficient in refining the microstructure and in improving the corrosion resistance of the composites. Hydroxyapatite also proved interesting for biomedical applications, e.g. as second phase in nitinol (i.e. NiTi intermetallic alloy) matrix composites [70].

3.4. Others

In view of their comparatively high cost, titanium matrix composites are somewhat less popular than Al-based or ferrous matrix composites. Again, carbides—particularly TiC [14, 15, 33, 71, 72], but also WC [73] or SiC [74]—have attracted a great interest as reinforcement in titanium matrix composites, with positive effect on their hardness and wear behaviour. TiB, TiB_2 [20, 75] and TiN [18] have also been used as alternative or in combination with TiC.

Copper (Cu) alloys are rather difficult to process by laser additive processes, due to their high reflectivity to laser light [26]. Nevertheless, the DLS of Cu matrix composites reinforced with Co/WC particles has been studied extensively, with the aim of combining the excellent thermal and electrical conductivity of the Cu matrix with the high strength and hardness of the Co/WC reinforcement [76, 77]. Rare earth oxides nanoparticles (CeO_2, La_2O_3) have been added in these Cu + Co/WC composites to favour microstructural refinement [78]. More recently, the direct addition of WC particles in Cu matrix composites processed by SLM has also been investigated [79].

Researches on Co-based matrix composites have been largely focused on the LC of composite reinforced with WC or other cermet particles, in view of applications necessitating a very high abrasion resistance as, e.g. in cutting tools [5, 80–82]. WC present a greater reactivity with Co alloys than with Ni alloys, and the partial dissolution of WC accompanied by secondary precipitation of finely dispersed complex carbides is commonly reported. A recent publication also reports on the LC of self-lubricating Co-based composite coatings with additions of TiC and CaF_2 [83].

4. Laser-induced in situ synthesis of MMCs

Reinforcement particles can also be synthesised in situ during the laser additive manufacturing of MMCs, either from a mixture of pure elemental powders [37, 84] or from ceramic particles that would decompose under the effect of laser irradiation and/or dissolve into the melt pool [2, 35, 38, 85]. The energy brought by the laser is used to fuse the metallic matrix powder and form new chemical compounds. Alternatively, the thermal energy brought by the laser may also trigger an exothermic chemical reaction that will not only produce new chemical compounds but also in itself generate enough thermal energy to propagate more chemical reactions [29, 86]. This process, known as self-propagating high-temperature synthesis, was originally used in combination with Selective Laser Sintering to produce Ni–Al intermetallics [39] and later extended to the synthesis of NiTi-matrix composites [40]. In this case, laser additive manufacturing can sometimes be pursued under a lower laser energy since the latent heat generated by the chemical reactions also contributes to increase the temperature of the melt pool [41, 86]. Besides, the in situ synthesis of MMCs has several other advantages: it allows for a better wetting and cohesion of ceramic particles with the metallic matrix, and for obtaining a fine and uniform distribution of the reinforcing phases [29].

The boundary between the ex situ fabrication and the in situ synthesis of MMCs is actually quite blurred. Indeed, in a few instances, the processing parameters for the fabrication of metal-ceramic composites and the size of the ceramic particles have been deliberately adjusted in such a way as to favour the partial dissolution of the ceramic and obtain reaction products with a well-tailored, fine and uniform distribution, thus taking advantage of the combined strengthening effect of the initial ceramic particles and of the secondary precipitates [2]. As a line of demarcation, the present section focuses on reviewing systematically works that involve a *conscious and deliberate* usage of chemical reactions in view of tailoring the microstructure and properties of MMCs.

4.1. In situ synthesis of aluminium matrix composites

Chang et al. [2] provides an interesting example of a careful and deliberate control of chemical reactions during the SLM of AlSi10Mg/SiC composites, making profit from the strong tendency of SiC to dissolve in Al and its alloys. As outlined previously (Section 3.1.1), the SLM processing parameters were adjusted in such a way as to favour the formation of mixed Al_4SiC_4 carbides over the potentially deleterious Al_4C_3. Moreover, the effect of the size of the SiC particles on the extent of secondary precipitation was investigated. It was found that the smallest average particle size of 5 μm gave way to the optimised distribution of refined Al_4SiC_4 precipitates and to the greatest improvement in hardness and wear behaviour.

Xu et al. [84] studied the in situ synthesis of Al-matrix composites reinforced with titanium diborides (TiB_2) from pure elemental Al, Ti and B powders. B powder was coated with Fe in order to avoid its burning during interaction with the laser. The resulting Al-matrix composites were reinforced with TiB_2 presenting a bimodal size distribution (i.e. with particles in size range 20–30 μm and smaller particles with a size < 500 nm). Besides, intermetallics Al_3Ti and Al_3Fe were also produced, and the wear resistance of the composites under low loads showed a marked improvement.

Dadbakhsh and Hao [87] attempted to produce Al matrix/oxides by SLM of Fe_2O_3 particles mixed with pure Al powder or various other Al alloys. This resulted in a very fine dispersion of Al_2O_3 and Al–Fe intermetallics with a size < 400 nm. Unfortunately, the mechanical behaviour of all the produced composites was dramatically compromised by an excessive volume fraction of porosities (i.e. at least 20 vol% or higher). Indeed, the oxygen introduced in the composite by the addition of Fe_2O_3 favoured the formation of an Al_2O_3 layer at the surface of the solidified material, thus preventing its proper wetting by the melt pool and a sufficient adhesion between successive layers which is a well-known issue in the SLM of Al alloys [88–90].

4.2. In situ synthesis of ferrous matrix composites

Ferrous matrix composites reinforced with various types of carbides have already been synthesised successfully using laser additive manufacturing techniques. Fe/TiC composites have been produced by LC, resulting in a significant increase in the hardness of the composite coating when compared to the substrate [37, 91]. Finely dispersed NbC were obtained from blended elemental powders during the LC of FeCrBSi matrix composites, with the addition of CeO_2 nanoparticles acting as heterogeneous nuclei enhancing the precipitation of NbC [57].

Song et al. [85] took advantage of the dissolution and reprecipitation of SiC micron-sized particles in pure Fe during SLM to produce Fe/SiC bulk nanocomposites. Similarly, Lo et al. [35] favoured the complete dissolution of WC particles and the reprecipitation of finely dispersed mixed carbides by selecting deliberately small (~1 μm) starting WC particles, with the aim of improving the cavitation erosion resistance of AISI316/WC composites.

As alternative to Fe/carbides composites, laser-cladded ferrous matrix composites reinforced with titanium diboride (TiB_2), exhibiting excellent hardness and wear resistance, were successfully synthesised from Fe–B and Fe–Ti alloyed powders [92]. In a study by Tan et al. [38], FeO_2 powder was added in a blend of pure elemental Fe, Ni, Cr, Al and graphite powders, with the aim of promoting the formation of finely distributed Al_2O_3 and various mixed carbides. A slight grain coarsening was observed in some specimens and interpreted as a consequence of the latent heat of reaction released by the exothermic formation of Al_2O_3.

4.3. In situ synthesis of MMCs in Ni- or Ti-based alloys

Ni-based composites reinforced with Cr_3C_2 [93], TiC and WC [86] carbides have been synthesised in situ from elemental powders. Lou et al. [93] investigated more particularly the influence of the Cr/C ratio on the resulting microstructure of NiCr/chromium carbides composites, with the aim of favouring the formation of Cr_3C_2 over mixed M_7C_3 and $M_{23}C_6$ carbides that are characterised by lower hardness and melting point in comparison with Cr_3C_2. Man et al. [86] reported on the successful laser-induced self-propagating high-temperature synthesis of TiC and WC in a NiAl intermetallic alloy. The latent heat of reaction released by the strongly exothermic formation of TiC was found to play an important role in promoting the formation of WC and of finely dispersed Al_3Ni_2 and Al_3Ni intermetallics, with a positive effect on the hardness of the composite clad layers.

Comparatively, the in situ synthesis of Ti-matrix composites reinforced with carbides has attracted less attention. Mixed (TiB+TiC)/Ti composites were synthesised in situ during the LC of blended boron carbide (B_4C) with pure elemental Ti and Al powder, resulting in excellent wear resistance [94]. Mixed (TiN+TiB)/Ti composites were synthesised during the LC of hexagonal boron nitride (h-BN) with Ti powders on a Ti-3Al-2V substrate, with the aim of combining the high hardness and Young's modulus of TiB with the enhanced high-temperature plastic behaviour of TiN [95]. High laser power should be preferred in order to ensure the complete dissociation of h-BN and an optimised formation of TiN and TiB, resulting in marked improvement of the wear resistance of the composite coating. In situ synthesised TiB/Ti composites have been the most popular so far among in situ Ti-based composites [96–99]. Earlier works [96, 97] focused on the obtention of a fine and uniform distribution of TiB precipitates in Ti-6Al-4V matrix composites. Recently, TiB particles have attracted a renewed interest in order to improve the wear resistance of Ti–Nb–Zr–Ta matrix composites for use in orthopaedic femoral implants, with the added advantage that additive manufacturing allows for the custom-designed fabrication of the implants [98, 99].

5. Synoptic table of laser additive manufacturing of MMCs and concluding remarks

The major characteristics of the different MMCs reviewed in Sections 3 and 4 are summarised in **Table 2**. In the following concluding remarks, special care is also taken to identify current trends and important fundamental issues in the laser additive manufacturing of MMCs.

Composite	Resulting microstructure	Remarks
Al-matrix with carbides		
AlSi10Mg + SiC by selective laser melting (SLM) / direct metal laser sintering (DMLS)	Partial dissolution of SiC and precipitation of finely dispersed Al_4SiC_4 [2] Dissolution of SiC and precipitation of Al_4C_3 [12]	Improved hardness and wear resistance [2] Moderate improvement in hardness [12]
Al-4,5Cu-3Mg + SiC by DMLS with pulsed laser	Partial dissolution of SiC and precipitation of Al_4C_3 [43, 44]	Extensive cracking due to high thermal gradients [43, 44]
Al-11 wt% Si + addition of Ti and WC by mixed wire/powder laser deposition	Ti added in the Al melt to promote the formation of TiC and inhibit the formation of Al_4C_3. Al–W and Al–Ti intermetallics are also observed [6]	–
Al-12 wt% Si + TiC by laser cladding	Very limited dissolution of TiC and formation of small amount of Ti_3SiC_2 [13]	Significant improvement in hardness and wear resistance [13]
AlSi10Mg + nano-TiC by SLM	AlSi10Mg and nano-TiC are pre-mixed to form a nanostructured powder [46–49]	Enhanced hardness and wear resistance [47]
Al-matrix with borides		
Al-12 wt% Si + TiB_2 by laser cladding	No sign of dissolution of TiB_2 [19]	Enhanced wear resistance [19]
Al + TiB_2 by in situ synthesis	Precipitation of finely dispersed TiB_2, Al_3Ti and Al_3Fe [84]	Enhanced wear resistance [84]
Al-matrix with oxides		
AlSi10Mg + nano-$MgAl_2O_4$ spinel by DMLS	Significant volume fraction of residual porosity, nanoparticles affect the solidification process [12]	Decrease in hardness [12]
Al + Fe_2O_3 by SLM	Excessive formation of Al_2O_3, high	–

Composite	Resulting microstructure	Remarks
	volume fraction of porosity [87]	
Fe-matrix with carbides		
Mild steel + SiC by laser cladding	Partial dissolution of SiC, formation of Fe_2Si [10]	Enhanced hardness, wear and corrosion resistance [10]
Stainless steel + SiC by laser cladding	Partial dissolution of SiC, precipitation of Fe_2Si, Fe_3Si, Fe_3Si_5, Cr_3C_2, Cr_7C_3 [11, 51–53]	Enhanced hardness and wear resistance [11, 51–53]
Fe + nano-SiC by DMLS	Dissolution of micron-size SiC and reprecipitation to form finely dispersed nano-SiC [85]	–
Stainless steel + WC by laser cladding	Partial dissolution of WC, precipitation of M_6C, $M_{23}C_6$, M_7C_3 [3, 35, 51, 54]	Enhanced hardness and wear resistance [3, 35, 51]
Stainless steel + TiC by laser cladding	Very little dissolution of TiC [51, 55]	–
Fe + TiC by in situ synthesis	Fine dispersion of TiC [37, 91]	Increase in hardness [37, 91]
Fe-36Ni "invar" + TiC by DMLS	Partial dissolution of TiC [56]	Loss of "invar" properties due to the enrichment of the FeNi matrix with Ti [56]
Stainless steel + Cr_3C_2 by laser cladding	Partial dissolution of Cr_3C_2 and precipitation of mixed M_7C_3 and $M_{23}C_6$ [3, 51]	Corrosion resistance is less compromised than with WC [3, 51]
Fe-matrix with oxides		
Stainless steel + Al_2O_3 by laser cladding	Al_2O_3 decomposes under laser irradiation [3, 34]	–
Stainless steel + Cr_2O_3 by laser cladding	–	Extensive cracking [51]
Fe-Ni-Cr-Al + Al_2O_3 by in situ synthesis	Finely dispersed Al_2O_3 and mixed carbides [92]	–
Fe-matrix with other particles		
Stainless steel + CrB_2 by laser cladding	Partial dissolution of CrB_2 and precipitation of finely dispersed CrB, Cr_2B and Fe_2B [51]	Enhanced hardness and cavitation erosion resistance [51]
FeCrBSi + NbC and nano-CeO_2	CeO_2 favours a fine dispersion NbC	–

Composite	Resulting microstructure	Remarks
by laser cladding	by acting as heterogeneous nuclei during solidification [57]	
Fe + TiB_2 by in situ synthesis	Finely dispersed TiB_2 [92]	–
Stainless steel + hydroxyapatite by SLM	–	Bioactive implants [58, 59]
Ni-matrix with carbides		
Ni + WC by laser cladding	Partial dissolution of WC and precipitation of various carbides [5, 60, 62]	–
NiCrBSi + WC by laser cladding	Partial dissolution of WC and precipitation of Ni–W intermetallics [63]	–
NiAl + WC and TiC by in situ synthesis	Ni-Al intermetallics are also observed [86]	Enhanced hardness [86]
Inconel 625 + Cr_3C_2 by laser cladding	Partial dissolution of Cr_3C_2 and precipitation of finely dispersed Cr-rich M_7C_3 [8]	Improved wear resistance [8]
NiCr + Cr_3C_2 by in situ synthesis	–	Cr/C ratio is varied to control the reaction products [93]
Inconel 625 or Inconel 690 + TiC by laser cladding	Minimal dissolution of TiC and precipitation of fine dendritic $M_{23}C_6$ [68]	–
Ni-matrix with other particles		
Ni + TiB_2 by laser cladding	Dispersed TiB_2 [26]	–
Ni + CeO_2 or Y_2O_3	CeO_2 and Y_2O_3 act as grain refiner [17, 69]	–
Nitinol + hydroxyapatite by SLS/SLM	–	Biomedical applications [70]
Ti-matrix composites		
Additions of TiC, WC, SiC, TiB TiB_2 or TiN	–	–
Ti + TiB and TiC by in situ synthesis	–	Excellent wear resistance [94]
Ti + TiB and TiN by in situ synthesis from h-BN	–	High laser power is recommended to ensure the full dissociation of h-BN [95]
Ti-6Al-4V + TiB by in situ synthesis	Fine dispersion of TiB [96, 97]	–

Composite	Resulting microstructure	Remarks
Ti-Nb-Zr-Ta + TiB	–	Enhanced wear resistance for use in femoral implant [98, 99]
Cu-matrix composites		
Cu + WC/Co by DMLS	–	Hardness and strength combined with good electrical and thermal conductivity [76–78]
Cu + WC by SLM	–	–
Co-matrix composites		
Co + WC	Partial dissolution of WC and precipitation of complex carbides [5, 80–82]	Applications in cutting tools [5, 80–82]
Co + TiC and CaF_2 by laser cladding	–	Self-lubricating MMCs [83]

Table 2. Synoptic table of representative MMCs fabricated by laser additive manufacturing.

Data from **Table 2** highlight the importance of carefully selecting the reinforcing phase of the composites in view of not only the desired properties and application but also considering the compatibility of the reinforcement with the metallic matrix:

- Carbides constitute the most popular reinforcement in combination with a wide range of metallic materials. However, in some occasions, they may give rise to deleterious reaction products such as the brittle Al_4C_3 in Al matrix composites or chromium carbides that may compromise the corrosion resistance of stainless steels. A careful optimisation of the laser additive manufacturing processing parameters may be necessary in order to avoid the formation of these undesirable reaction products.

- Nitrides and borides are interesting alternative to carbides, in view of their generally good chemical stability and of their excellent mechanical properties.

- As a general rule, oxides appear much less promising for use as reinforcing phase in MMCs fabricated by laser additive manufacturing. Oxide particles may easily decompose during laser irradiation, and they present a strong tendency to dissolve in the melt pool and reprecipitate as a detrimental oxide skin that would compromise the good cohesion of successive additive-manufactured layers. A notable exception is the addition of rare earth oxide nanoparticles that act as very potent microstructural refiner, with beneficial effects on the mechanical and corrosion properties of the MMCs.

- Nanocomposites with enhanced properties have been successfully fabricated by laser additive manufacturing. However, specific procedures are necessary in order to avoid the processability issues related with the high tendency of nanoparticles to agglomerate and with their low flowability. For the production of ex situ MMCs, a nanostructured powder should be prepared by carefully mixing the nanoparticles with the metallic powder used for the matrix. Laser-induced in situ synthesis of MMCs is a very interesting alternative, allowing for a refined and uniform distribution of nanoparticles.

Author details

Anne I. Mertens* and Jacqueline Lecomte-Beckers

*Address all correspondence to: anne.mertens@ulg.ac.be

University of Liege, Aerospace and Mechanical Engineering, Metallic Materials Science, Liège, Belgium

References

[1] Kainer K.U. Basics of Metal Matrix Composites. In: Kainer K.U., editor. Metal Matrix Composites. Custom-made Materials for Automotive and Aerospace Engineering. Weinheim: Wiley-VCH Verlag; 2006. p. 1-54.

[2] Chang F., Gu D., Dai D., Yuan P. Selective laser melting of in-situ Al_4SiC_4 + SiC hybrid reinforced Al matrix composites: Influence of starting particle size. Surface & Coatings Technology. 2015;272:15-24. DOI: 10.1016/j.surfcoat.2015.04.029

[3] Betts J.C., Mordike B.L., Grech M. Characterisation, wear and corrosion testing of laser-deposited AISI 316 reinforced with ceramic particles. Surface Engineering. 2010;26(1-2): 21-29. DOI: 10.1179/174329409X433920

[4] Jankauskas V., Antonov M., Varnauskas V., Skirkus R., Goljandin D. Effect of WC grain size and content on low stress abrasive wear of manual arc welded hardfacing with low-carbon or stainless steel matrix. Wear. 2015;328-329:378-390. DOI: 10.1016/j.wear. 2015.02.063

[5] Przybylowicz J., Kusinski J. Structure of laser cladded tungsten carbide composite coatings. Journal of Materials Processing Technology. 2001;109:154-160.

[6] Li F., Gao Z., Zhang Y., Chen Y. Alloying effect of titanium on WCp/Al composite fabricated by coincident wire-powder laser deposition. Materials and Design. 2016;93:370-378. DOI: 10.1016/j.matdes.2015.12.171

[7] Farayibi P.K., Folkes J., Clare A., Oyelola O. Cladding of pre-blended Ti-6Al-4V and WC powder for wear resistant applications. Surface & Coatings Technology. 2011;206:372-377. DOI: 10.1016/j.surfcoat.2011.07.033

[8] Verdi D., Garrido M.A., Múnez C.J., Poza P. Cr_3C_2 incorporation into an Inconel 625 laser cladded coating: Effects on matrix microstructure, mechanical properties and local

scratch resistance. Materials and Design. 2015;67:20-27. DOI: 10.1016/j.matdes.2014.10.086

[9] Abenojar J., Velasco F., Torralba J.M., Bas J.A., Calero J.A., Marcé R. Reinforcing 316L stainless steel with intermetallic and carbide particles. Materials Science and Engineering A. 2002;335:1-5.

[10] Dutta Majumdar J., Ramesh Chandra B., Nath A.K., Manna I. Studies on compositionally graded silicon carbide dispersed composite surface on mild steel developed by laser surface cladding. Journal of Materials Processing Technology. 2008;203:505-512. DOI: 10.1016/j.jmatprotec.2007.10.056

[11] Dutta Majumdar J., Kumar A., Li L. Direct laser cladding of SiC dispersed AISI 316L stainless steel. Tribology International. 2009;42:750-753. DOI: 10.1016/j.triboint.2008.10.016

[12] Manfredi D., Calignano F., Krishnan M., Canali R., Ambrosio E.P., Biamino S. et al. Additive Manufacturing of Al Alloys and Aluminium Matrix Composites (AMCs). In: Monteiro W.A., editor. Light Metal Alloys Applications. Intech; 2014. p. 3-34. DOI: 10.5772/58534

[13] Anandkumar R., Almeida A., Vilar R. Microstructure and sliding wear resistance of an Al-12wt.% Si/TiC laser clad coating. Wear. 2012;282-283:31-39. DOI: 10.1016/j.wear.2012.01.022

[14] Candel J.J., Amigó V., Ramos J.A., Busquets D. Sliding wear resistance of TiCp reinforced titanium composite coating produced by laser cladding. Surface & Coatings Technology. 2010;204:3161-3166. DOI: 10.1016/j.surfcoat.2010.02.070

[15] Mahamood R.M., Akinlabi E.T. Laser metal deposition of functionally graded Ti6Al4V/TiC. Materials and Design. 2015;84:402-410. DOI: 10.1016/j.matdes.2015.06.135

[16] Xu P.Q., Gong H.Y., Xu G.X., He J.P., Yu Z.S. Study on microstructure and properties of Ni-based alloy/Y_2O_3-deposited metals by laser cladding. Journal of Materials Science. 2008;43:1559-1567. DOI: 10.1007/s10853-007-2339-6

[17] Wang H., Zuo D., Chen G., Sun G., Li X., Cheng X. Hot corrosion behaviour of low Al NiCoCrAlY cladded coatings reinforced by nano-particles on a Ni-base super alloy. Corrosion Science. 2010;52:3561-3567. DOI: 10.1016/j.corsci.2010.07.011

[18] Balla V.K., Bhat A., Bose S., Bandyopadhyay A. Laser processed TiN reinforced Ti6Al4V composite coatings. Journal of the Mechanical Behavior of Biomedical Materials. 2012;6:9-20. DOI: 10.1016/j.jmbbm.2011.09.007

[19] Anandkumar R., Almeida A., Vilar R. Wear behavior of Al-12Si/TiB_2 coatings produced by laser cladding. Surface & Coatings Technology. 2011;205:3824-3832. DOI: 10.1016/j.surfcoat.2011.01.048

[20] Diao Y., Zhang K. Microstructure and corrosion resistance of TC2 Ti alloy by laser cladding with Ti/TiC/TiB_2 powders. Applied Surface Science. 2015;352:163-168. DOI: 10.1016/j.apsusc.2015.04.030

[21] Spanou S., Kontos A.I., Siokou A., Kontos A.G., Vaenas N., Falaras P. et al. Self cleaning behaviour of Ni/nano-TiO_2 metal matrix composites. Electrochimica Acta. 2013;105:324-332. DOI: 10.1016/j.electacta.2013.04.174

[22] Ferguson J.B., Schultz B.F., Rohatgi P.K. Self-healing metals and metal matrix composites. JOM. 2014;66(6):866-871. DOI: 10.1007/s11837-014-0912-4

[23] Moghadam A.D., Schultz B.F., Ferguson J.B., Omrani E., Rohatgi P.K., Gupta N. Functional metal matrix composites: Self-lubricating, self-healing, and nanocomposites - an outlook. JOM. 2014;66(6):872-881. DOI: 10.1007/s11837-014-0498-5

[24] Mahathanabodee S., Palathai T., Raadnui S., Tongsri R., Sombatsompop N. Dry sliding wear behavior of SS316L containing h-BN and MoS_2 solid lubricants. Wear. 2014;316:37-48. DOI: 10.1016/j.wear.2014.04.015

[25] Liu X.B., Shi S.H., Guo J., Fu G.Y., Wang M.D. Microstructure and wear behavior of gamma/Al_4C_3/TiC/CaF_2 composite coating on gamma-TiAl intermetalllic alloy prepared by Nd: YAG laser cladding. Applied Surface Science. 2009;255:5662-5668. DOI: 10.1016/j.apsusc.2008.11.023

[26] Yan H., Zhang P., Yu Z., Lu Q., Yang S., Li C. Microstructure and tribological properties of laser-clad Ni–Cr/TiB_2 composite coatings on copper with the addition of CaF_2. Surface & Coatings Technology. 2012;206:4046-4053. DOI: 10.1016/j.surfcoat.2012.03.086

[27] Mertens A., Simar A., Delannay F. C fibres–Mg matrix composites produced by squeeze casting and friction stir processing: Microstructure & mechanical behaviour. Materials Science Forum. 2012;706-709:1221-1226. DOI: 10.4028/www.scientific.net/MSF.706-709.1221

[28] Arora H.S., Singh H., Dhindaw B.K. Composite fabrication using friction stir processing - a review. International journal of Advanced Manufacturing Technologies. 2012;61:1043-1055. DOI: 10.1007/s00170-011-3758-8

[29] Kumar S., Kruth J.P. Composites by rapid prototyping technology. Materials and Design. 2010;31:850-856. DOI: 10.1016/j.matdes.2009.07.045

[30] Verlee B., Dormal T., Lecomte-Beckers J. Density and porosity control of sintered 316L stainless steel parts produced by additive manufacturing. Powder Metallurgy. 2012;55(4):260-267. DOI: 10.1179/003289912z.00000000082

[31] Quan Z., Wu A., Keefe M., Qin X., Yu J., Suhr J. et al. Additive manufacturing of multidirectional preforms for composites: Opportunities and challenges. Materials Today. 2015;18(9):503-512. DOI: 10.1016/j.mattod.2015.05.001

[32] Frazier W.E. Metal additive manufacturing: A review. Journal of Materials Engineering and Performance. 2014;23(6):1917-1928. DOI: 10.1007/s11665-014-0958-z

[33] Wang F., Mei J., Jiang H., Wu X. Laser fabrication of Ti6Al4V/TiC composites using simultaneous powder and wire feed. Materials Science and Engineering A. 2007;445-446:461-466. DOI: 10.1016/j.msea.2006.09.093

[34] Xu P., Lin C., Zhou C., Yi X. Wear and corrosion resistance of laser cladding AISI304 stainless steel/Al_2O_3 composite coatings. Surface & Coatings Technology. 2014;238:9-14. DOI: 10.1016/j.surfcoat.2013.10.028

[35] Lo K.H., Cheng F.T., Kwok C.T., Man H.C. Improvement of cavitation erosion resistance of AISI 316 stainless steel by laser surface alloying using fine WC powder. Surface & Coatings Technology. 2003;165:258-267.

[36] L'Hoest T. Production of metal/ceramic composite coatings by laser cladding [thesis]. Liège (Belgium): University of Liege; 2015. p. 125.

[37] Emamian A., Alimardani M., Khajepour A. Correlation between temperature distribution and in situ formed microstructure of Fe–TiC deposited on carbon steel using laser cladding. Applied Surface Science. 2012;258:9025-9031. DOI: 10.1016/j.apsusc.2012.05.143

[38] Tan H., Luo Z., Li Y., Yan F., Duan R., Huang Y. Effect of strengthening particles on the dry sliding wear behavior of Al_2O_3-M_7C_3/Fe metal matrix composite coatings produced by laser cladding. Wear. 2015;324-325:36-44. DOI: 10.1016/j.wear.2014.11.023

[39] Shishkovskii I.V., Makarenko A.G., Petrov A.L. Conditions for SHS of intermetallic compounds with selective laser sintering of powdered compositions. Combustion, Explosion and Shock Waves. 1999;35(2):166-170.

[40] Shishkovsky I.V., Tarasova E.Yu., Zhuvarel L.V., Petrov A.L. The synthesis of a biocomposite based on nickel titanium and hydroxyapatite under selective laser sintering conditions. Technical Physics Letters. 2001;27(3):211-213.

[41] Shishkovsky I.V. Laser-Controlled Intermetallics Synthesis during Surface Cladding. In: Lawrence J., Waugh D.G, editors. Laser Surface Engineering. Woodhead; 2015. p. 237-286. DOI: 10.1016/B978-1-78242-074-3.00011-8

[42] Lee J.C., Byun J.Y., Park S.B., Lee H.I. Prediction of Si contents to suppress the formation of Al_4C_3 in the SiCp/Al composite. Acta Materialia. 1998;46(5):1771-1780.

[43] Ghosh S.K., Saha P., Kishore S. Influence of size and volume fraction of SiC particulates on properties of ex situ reinforced Al-4.5Cu-3Mg metal matrix composite prepared by direct metal laser sintering process. Materials Science and Engineering A. 2010;527:4694-4701. DOI: 10.1016/j.msea.2010.03.108

[44] Ghosh S.K., Saha P. Crack and wear behaviour of SiC particulate reinforced aluminium based metal matrix composite fabricated by direct metal laser sintering process. Materials and Design. 2011;32:139-145. DOI: 10.1016/j.matdes.2010.06.020

[45] Kadolkar P.B., Watkins T.R., De Hosson J.Th.M., Kooi B.J., Dahotre N.B. State of residual stress in laser-deposited ceramic composite coatings on aluminium alloys. Acta Materialia. 2007;55:1203-1214. DOI: 10.1016/j.actamat.2006.07.049

[46] Gu D., Wang H., Chang F., Dai D., Yuan P., Hagedorn Y.C. et al. Selective laser melting additive manufacturing of TiC/AlSi10Mg bulk-form nanocomposites with tailored microstructures and properties. Physics Procedia. 2014;56:108-116. DOI: 10.1016/j.phpro.2014.08.153

[47] Gu D., Wang H., Dai D., Chang F., Meiners W., Hagedorn Y.C. et al. Densification behavior, microstructure evolution, and wear property of TiC nanoparticle reinforced AlSi10Mg bulk-form nanocomposites prepared by selective laser melting. Journal of Laser Applications. 2015;27:S17003. DOI: 10.2351/1.4870877

[48] Dai D., Gu D. Tailoring surface quality through mass and momentum transfer modeling using a volume fluid method in selective laser melting of TiC/AlSi10Mg powder. International Journal of Machine Tools & Manufacture. 2015;88:95-107. DOI: 10.1016/j.ijmachtools.2014.09.010

[49] Gu D., Wang H., Dai D., Yuan P., Meiners W., Poprawe R. Rapid fabrication of Al-based bulk-form nanocomposites with novel reinforcement and enhanced performance by selective Laser melting. Scripta Materialia. 2015;96:25-28. DOI: 10.1016/j.scriptamat.2014.10.011

[50] Thawari G., Sundararajan G., Joshi S.V. Laser surface alloying of medium carbon steel with SiC(p). Thin Solid Films. 2003;423:41-53.

[51] Cheng F.T., Kwok C.T., Man H.C. Laser surfacing of S31603 stainless steel with engineering ceramics for cavitation erosion resistance. Surface & Coatings Technology. 2001;139:14-24.

[52] Abbas G., Ghazanfar U. Two-body abrasive wear studies of laser produced stainless steel and stainless steel + SiC composite clads. Wear. 2005;258:258-264. DOI: 10.1016/j.wear.2004.09.036

[53] Dutta Majumdar J., Li L. Studies on Direct laser Cladding of SiC Dispersed AISI316L Stainless Steel. Metallurgical and Materials Transactions A. 2009;40:3001-3008. DOI: 10.1007/s11661-009-0018-8

[54] Babu S.S., Martukanitz R.P., Parks K.D., David S.A. Toward Prediction of Microstructural Evolution during Laser Surface Alloying. Metallurgical and Materials Transactions A. 2002;33:1189-1200.

[55] Das K., Bandyopadhyay T.K., Das S. A review on the various synthesis routes of TiC reinforced ferrous based composites. Journal of Materials Science. 2002;37:3881-3892.

[56] Gaard A., Krakhmalev P., Bergstrom J. Microstructural characterization and wear behavior of (Fe,Ni)-TiC MMC prepared by DMLS. Journal of Alloys and Compounds. 2006;421:166-171. DOI: 10.1016/j.jallcom.2005.09.084

[57] Li Q., Lei Y., Fu H. Laser cladding in-situ NbC particle reinforced Fe-based composite coatings with rare earth oxide addition. Surface & Coatings Technology. 2014;239:102-107. DOI: 10.1016/j.surfcoat.2013.11.026

[58] Hao L., Dadbakhsh S., Seaman O., Felstead M. Selective laser melting of a stainless steel and hydroxyapatite composite for load-bearing implant development. Journal of Materials Processing Technology. 2009;209:5793-5801. DOI: 10.1016/j.jmatprotec.2009.06.012

[59] Wei Q., Li S., Han Q., Li W., Cheng L., Hao L. et al. Selective laser melting of stainless steel/nano-hydroxyapatite composites for medical applications: Microstructure, element distribution, crack and mechanical properties. Journal of Materials Processing Technology. 2015;222:444-453. DOI: 10.1016/j.jmatprotec.2015.02.010

[60] Cheng F.T., Kwok C.T., Man H.C. Cavitation erosion resistance of stainless steel laser-clad with WC-reinforced MMC. Materials Letters. 2002;57:969-974.

[61] Wu P., Zhou C.Z., Tang X.N. Microstructural characterization and wear behavior of laser cladded nickel-based and tungsten carbide composite coatings. Surface & Coatings Technology. 2003;166:84-88.

[62] Sahour M.C., Bahloul A., Vannes A.B. Characteristics of the laser clad metal made with powder mixture of Ni-based alloy and tungsten carbide. International Journal of Material Forming. 2008;Suppl 1:1379-1382. DOI: 10.1007/s122289-008-0121-1

[63] Guo C., Chen J., Zhou J., Zhao J., Wang L., Yu Y. et al. Effects of WC–Ni content on microstructure and wear resistance of laser cladding Ni-based alloys coating. Surface & Coatings Technology. 2012;206:2064-2071. DOI: 10.1016/j.surfcoat.2011.06.005

[64] Van Acker K., Vanhoyweghen D., Persoons R., Vangrunderbeek J. Influence of tungsten carbide particle size and distribution on the wear resistance of laser clad WC/Ni coatings. Wear. 2005;258:194-202. DOI: 10.1016/j.wear.2004.09.041

[65] St-Georges L. Development and characterization of composite Ni–Cr + WC laser cladding. Wear. 2007;263:562-566. DOI: 10.1016/j.wear.2007.02.023

[66] Huang F., Jiang Z., Liu X., Lian J., Chen L. Microstructure and properties of thin wall by laser cladding forming. Journal of Materials Processing Technology. 2009;209:4970-4976. DOI: 10.1016/j.jmatprotec.2009.01.019

[67] Cooper D.E., Blundell N., Maggs S., Gibbons G.J. Additive layer manufacture of Inconel 625 metal matrix composites, reinforcement material evaluation. Journal of Materials Processing Technology. 2013;213:2191-2200. DOI: 10.1016/j.matprotec.2013.06.021

[68] Wilson J.M., Shin Y.C. Microstructure and wear properties of laser-deposited functionally graded Inconel 690 reinforced with TiC. Surface & Coatings Technology. 2012;207:517-522. DOI: 10.1016/j.surfcoat.2012.07.058

[69] Pérez A.T., Montealegre M.A., Arias J.L., Castro G., Rey P., González M. et al. Laser cladding of nickel alloy with ceramic nanopowder on steel as coating in corrosive media. Materials Science Forum. 2012;706-709:2552-2558. DOI: 10.4028/www.scientific.net/MSF/.706-709.2552

[70] Shishkovskii I.V., Yadroitsev I.A., Smurov I.Y. Selective laser sintering/melting of nitinol-hydroxyapatite composite for medical applications. Powder Metallurgy and Metal Ceramics. 2011;50(5-6):275-283.

[71] Sampedro J., Pérez I., Carcel B., Ramos J.A., Amigo V. Laser cladding of TiC for better titanium components. Physics Procedia. 2011;12:313-322. DOI: 10.1016/j.phpro.2011.03.040

[72] Gu D., Wang H., Zhang G. Selective laser melting additive manufacturing of Ti-based nanocomposites: The role of nanopowder. Metallurgical and Materials Transactions A. 2014;45:464-476. DOI: 10.1007/s11661-013-1968-4

[73] Farayibi P.K., Folkes J.A., Clare A.T. Laser deposition of Ti-6Al-4V wire with WC powder for functionally graded components. Materials and Manufacturing Processes. 2013;28(5):514-518. DOI: 10.1080/10426914.2012.718477

[74] Pei Y.T., Ocelik V., De Hosson J.Th.M. SiCp/Ti6Al4V functionally graded materials produced by laser melt injection. Acta Materialia. 2002;50:2035-2051.

[75] Galvan D., Ocelik V., Pei Y., Kooi B.J., De Hosson J.Th.M., Ramous E. Microstructure and properties of TiB/Ti-6Al-4V coatings produced with laser treatments. Journal of Materials Engineering and Performance. 2004;13:406-412. DOI: 10.1361/10599490419919

[76] Gu D., Shen Y. Processing and microstructure of submicron WC–Co particulate reinforced Cu matrix composites prepared by direct laser sintering. Materials Science and Engineering A. 2006;435-436:54-61. DOI: 10.1016/j.msea.2006.07.105

[77] Gu D., Shen Y. Direct laser sintered WC–Co/Cu nanocomposites. Applied Surface Science. 2008;254:3971-3978. DOI: 10.1016/j.apsusc.2007.12.028

[78] Gu D., Shen Y., Zhao L., Xiao J., Wu P., Zhu Y. Effect of rare earth oxide addition on microstructures of ultra-fine WC–Co particulate reinforced Cu matrix composites prepared by direct laser sintering. Materials Science and Engineering A. 2007;445-446:316-322. DOI: 10.1016/j.msea.2006.09.057

[79] Dai D., Gu D. Thermal behavior and densification mechanism during selective laser melting of copper matrix composites: Simulation and experiments. Materials and Design. 2014;55:482-491. DOI: 10.1016/j.matdes.2013.10.006

[80] Xu G., Kutsuna M., Liu Z., Sun L. Characteristic behaviours of clad layer by a multi-layer laser cladding with powder mixture of Stellite-6 and tungsten carbide. Surface & Coatings Technology. 2006;201:3385-3392. DOI: 10.1016/j.surfcoat.2006.07.210

[81] Nurminen J., Näkki J., Vuoristo P. Microstructure and properties of hard and wear resistant MMC coatings deposited by laser cladding. International Journal of Refractory Metals & Hard Materials. 2009;27:472-478. DOI: 10.1016/j.ijrmhm.2008.10.008

[82] Doubenskaia M., Gilmutdinov, Nagulin K.Y. Laser cladding of metal matrix composites reinforced by cermet inclusions for dry friction applications at ambient and elevated temperatures. Surface & Coatings Technology. 2015;276:696-703. DOI: 10.1016/j.surfcoat.2015.05.044

[83] Yan H., Zhang J., Zhang P., Yu Z., Li C., Xu P. et al. Laser cladding of Co-based alloy/TiC/CaF_2 self-lubricating composite coatings on copper for continuous casting mold. Surface & Coatings Technology. 2013;232:362-369. DOI: 10.1016/j.surfcoat.2013.05.036

[84] Xu J., Liu W. Wear characteristic of in situ synthetic TiB_2 particulate-reinforced Al matrix composite formed by laser cladding. Wear. 2006;260:486-492. DOI: 10.1016/j.wear.2005.03.032

[85] Song B., Dong S., Coddet C. Rapid in situ fabrication of Fe/SiC bulk nanocomposites by selective laser melting directly from a mixed powder of microsized Fe and SiC. Scripta Materialia. 2014;75:90-93. DOI: 10.1016/j.scriptamat.2013.11.031

[86] Man H.C., Yang Y.Q., Lee W.B. Laser induced reaction synthesis of TiC+WC reinforced metal matrix composites coatings on Al 6061. Surface & Coatings Technology. 2004;185:74-80. DOI: 10.1016/j.surfcoat.2003.10.132

[87] Dadbakhsh S., Hao L. Effect of Al alloys on selective laser melting behaviour and microstructure of in situ formed particle reinforced composites. Journal of Alloys and Compounds. 2012;541:328-334. DOI: 10.1016/j.allcom.2012.06.097

[88] Louvis E., Fox P., Sutcliffe C.J. Selective laser melting of aluminium components. Journal of Materials Processing Technology. 2011;211:275-284. DOI: 10.1016/j.matprotec.2010.09.019

[89] Olakanmi E.O. Selective laser sintering/melting (SLS/SLM) of pure Al, Al–Mg, and Al–Si powders: Effect of processing conditions and powder properties. Journal of Materials Processing Technology. 2013;213:1387-1405. DOI: 10.1016/j.matprotec.2013.03.009

[90] Aboulkhair N.T., Everitt N.M., Ashcroft I., Tuck C. Reducing porosity in AlSi10Mg parts processed by selective laser melting. Additive Manufacturing. 2014;1-4:77-86. DOI: 10.1016/J.addma.2014.08.001

[91] Emamian A., Alimardani M., Khajepour A. Effect of cooling rate and laser process parameters on additive manufactured Fe–Ti–C metal matrix composites microstruc-

ture and carbide morphology. Journal of Manufacturing Processes. 2014;16:511-517. DOI: 10.1016/j.mapro.2014.07.002

[92] Du B., Zou Z., Wang X., Qu S. Laser cladding of in situ TiB_2/Fe composite coating on steel. Applied Surface Science. 2008;254:6489-6494. DOI: 10.1016/j.apsusc.2008.04.051

[93] Lou D., Liu D., He C., Bennett P., Chen L., Yang Q. et al. Effect of Cr/C ratio on microstructure and corrosion performance of Cr_3C_2-NiCr composite fabricated by laser processing. Journal of Materials Engineering and Performance. 2016;25:312-319. DOI: 10.1007/s11665-015-1843-0

[94] Li J., Yu Z., Wang H. Wear behaviors of an (TiB+TiC)/Ti composite coating fabricated on Ti6Al4V by laser cladding. Thin Solid Films. 2011;519:4804-4808. DOI: 10.1016/j.tsf. 2011.01.034

[95] Li M., Huang J., Zhu Y.Y., Li Z.G. Effect of heat input on the microstructure of in-situ synthesized TiN-TiB/Ti based composite coating by laser cladding. Surface & Coatings Technology. 2012;206:4021-4026. DOI: 10.1016/j.surfcoat.2012.03.082

[96] Banerjee R., Collins P.C., Genç A., Fraser H.L. Direct laser deposition of in situ Ti-6Al-4V-TiB composites. Materials Science and Engineering A. 2003;358:343-349. DOI: 10.1016/S0921-5093(03)00299-5

[97] Banerjee R., Genç A., Hill D., Collins P.C., Fraser H.L. Nanoscale TiB precipitates in laser deposited Ti-matrix composites. Scripta Materialia. 2005;53:1433-1437. DOI: 10.1016/j.scriptamat.2005.08.014

[98] Samuel S., Nag S., Scharf T.W., Banerjee R. Wear resistance of laser-deposited boride reinforced Ti–Nb–Zr–Ta alloy composites for orthopedic implants. Materials Science and Engineering C. 2008;28:414-420. DOI: 10.1016/j.msec.2007.04.029

[99] Nag S., Samuel S., Puthucode A., Banerjee R. Characterisation of novel borides in Ti–Nb–Zr–Ta+2B metal matrix composites. Materials Characterization. 2009;60:106-113. DOI: 10.1016/j.matchar.2008.07.011

Chapter 10

Metal Powder Additive Manufacturing

Anatoliy Popovich and Vadim Sufiiarov

Additional information is available at the end of the chapter

http://dx.doi.org/10.5772/63337

Abstract

The beginning of the chapter is devoted to methods of receiving of metal powders—initial components for metal additive manufacturing. Initial materials are very important part of manufacturing, because their quality has an influence on stability of production process and quality of final product. There are various methods of metal powder synthesis. They may be separated conventionally on physical–chemical and mechanical ones. The physical–chemical methods are associated with physical and chemical transformations, and chemical composition and structure of the final product (metal powder) significantly differ from raw materials. The mechanical methods include various types of milling processes and jet dispersion melts by high pressure of gas or liquid (atomization). It is shown that the typical methods of quantitative estimation of powdered materials and some parameters for alloys that already were used in additive technologies. The next theme of the chapter is a review of additive technologies, initial materials for that is metal powders. At this moment, there are three main technologies that have found wide use for the production of metal parts from metal powders: binder jetting, directed energy deposition, and powder bed fusion. Each of them has unique peculiarity, advantages, and limitations that will be presented in the chapter.

Keywords: powder metallurgy, atomization, additive manufacturing, additive technologies, selective laser melting

1. Introduction

Additive manufacturing technologies are currently one of the fastest growing manufacturing processes. The technologies provide engineers an innovative approach for design and manufacture of parts. They substantially reduce the amount of post-processing and improve product quality by producing parts with form the closest to computer model data. All the variety of additive technologies are available in the annual report [1].

An important part of additive manufacturing of metal parts is the initial material. There are different approaches of additive manufacturing, which use different types of initial materials, and the most popular technologies, such as selective laser or electron beam melting, laser cladding, and binder jetting, use initial material in the powdered form [2–4], but there are also technologies which use initial material in sheet or wire form [5, 6].

In the chapter, the state of art of metal powder based on additive manufacturing will be presented. The chapter considers three main themes—metal powders, properties of metal powders, additive technologies, and properties of metal parts. It will be shown the methods for mass production of metal powders for additive manufacturing technologies, descriptions on characterization of powder properties and microstructure and mechanical properties of metal samples.

2. Technologies of metal powder production

There are various technologies for mass metal powder production, and it should be marked that in the chapter will be shown technologies available for mass powder production and will not be considered such technologies as sol–gel, chemical vapor deposition, and physical vapor deposition that allow to receive nanosized powders with unique properties but not applicable in additive manufacturing at this moment. One of the main requirements for using of metal powder in additive manufacturing and receiving reliable and repeatable results is a spherical form of particles. Some technologies allow to produce a spherical or near to spherical powder shape directly after synthesis of powder, whereas the other technologies require a further processing to achieve the desired particles shape. Technologies for the production of metal powder conventionally are separated on base of the following methods: physical–chemical and mechanical ones. The physical–chemical methods are associated with physical and chemical transformations, chemical composition, and structure of the final product (metal powder) and significantly differ from raw materials. The mechanical methods include various types of milling processes and jet dispersion melts by high pressure of gas or liquid (also known as atomization).

2.1. Mechanical methods

Atomization is the most widespread technology for the mass production of metal powders for additive manufacturing. There are various methods; the most popular is a gas atomization process; similar to water atomization technology, another one is a plasma atomization, also known as rotating electrode atomization; and less popular is a centrifugal atomization.

The main principle of all atomization technologies is a disintegration (dispersion) of a thin stream of molten metal by subjecting it with impact of gas, high pressure of water, plasma, rotating forces etc. During this impact, molten metal is divided on small droplets, which rapidly crystallize in flight before they reach atomizer walls.

Gas atomization—at this moment, this is the main process for producing of metal powders for additive manufacturing. The process steps involved into the production of metal powders are melting, atomizing, and solidifying of the respective metals and alloys. Gas atomizers are usually equipped with a furnace for melting under vacuum or rarely under protective atmosphere, with feeders of liquid alloy with nozzles in atomizing chamber, where a thin flow of the melted alloy dispersed on small droplets by high pressure of inert gas, and the droplets solidify during the flight in atomizing chamber. Powders produced by gas atomization have a spherical shape, high cleanliness, fine, and homogeneous microstructure (thanks to rapid solidification).

One of the European leaders in producing of equipment for gas atomization is the German company ALD vacuum technologies GmbH. The company offers different modifications of gas atomizer for producing of different alloys, which allow to produce powders and a wide range of metals and alloys. Two main modifications are VIGA and EIGA. The first one is decrypted as a vacuum induction melting combined with inert gas atomization, and this is the most popular system that allows to produce powders of nonreactive metals and their alloys. The second one is decrypted as electrode induction melting gas atomization, and this system uses the high-reactive metals and alloys such as titanium for powder production. The other modifications are less popular and have been used in special cases:

– Plasma melting induction guiding gas atomization (PIGA) uses a plasma burner instead of melting induction and water-cooled copper crucible. This system usually used for the production of ceramic-free and reactive high-melting alloys;

– Electroslag remelting–cold wall induction guiding (ESR-CIG) was especially developed for the production of high performance of superalloys. It uses the so-called "triple melt process" for reaching the highest level of cleanliness and chemical homogeneity of powder. This system uses water-cooled copper crucible, same as in PIGA, and raw material in form of an electrode, as in EIGA;

– Vacuum induction melting based on the cold wall crucible melting technology combined with inert gas atomization (VIGA-CC) was developed for the production of reactive alloys, which are difficult to produce in electrode form (for example, brittle intermetallic TiAl alloys) and use water-cooled copper crucible with a bottom pouring system [7].

Powders obtained by gas atomization process usually have a spherical or near to spherical shape and have particle sizes, which mostly can be used in additive technologies. It should be noted that particle size distribution has a strong dependence on the type of atomized alloy and used system.

Water atomization is similar to gas atomization process, but instead of gas, it uses high pressure of water steam as atomizing medium. The water atomization is used mostly for the production of powders, unreactive materials such as steels. Due to higher cooling rates in comparing to the gas atomization, particles have irregular shapes. The main advantage of water atomization consists in the fact that it is less expensive process than the other types of atomization; disadvantage is in the limitations of purity, especially for metals and alloys inclined to oxidation.

Another relatively non-expensive process is a compressed air atomization. This process also is used to produce unreactive materials, and particles shapes have many defects such as satellites, internal porosity etc.

Plasma atomization is a relatively new process, which was developed for production of high-purity powders of reactive metals and alloys with high melting point such as titanium, zirconium, tantalum etc. Plasma atomization allows to produce fine particle distribution powders with highly spherical particles shape and low content of oxygen. The initial material for plasma atomization process is a metal wire. Wire feedstock is fed into a plasma torches that disperse wire into droplets with subsequent solidification in powder form. Particle size distribution of powder produced by plasma atomization is 0–200 μm.

The use of a wire has advantages over the typical gas atomization process. The most significant advantage consists in the fact that the metal feedstock, and more importantly the melt, does not come into contact with cold solid surfaces. This is another approach in comparison with the use of cooling crucible to receive high-purity powders. The first production step is a wire feeding, and the speed of the wire should be monitored in order to control and adjust the resulting particle size distribution. The low flow rate of argon is used because of using argon plasma as the atomizing medium as well as heat source, since the heated gas has a higher velocity, and thus, a stronger atomization force is applied. Additionally, the use of a hot atomizing gas instead of a cold one prevents the particles the rapidly freezing of particles together into irregular shapes. The use of plasma as a heating source enables to reach a high superheat and the result of cooling ensures to complete spheroidization. Powder collection is occurred with a typical cyclonic device, and the powder is carefully passivated to ensure the safe manipulation in the open air [8].

There is a limitation for plasma atomization technology—initial material has to be flexible enough to get it in wire feedstock, so it is impossible to atomize materials that can not to be produced in a wire form. There are two companies in Canada, which use plasma atomization as the main process for powder production: AP&C (ex Raymor) and PyroGenesis, both companies produce powders with the focus on application of additive manufacturing.

2.1.1. Centrifugal atomization

The other types of atomization processes comprise a number of centrifugal atomization processes. There exist several schemes of using centrifugal forces for dispersing of molten metal; however, two types of such processes are more popular. The first type is a rotating electrode process (REP); a metal electrode is rotated with high velocity; and the free end is melted with an arc between the metal electrode and the tungsten electrode; if a plasma arc is involved, the process is known as plasma rotating electrode process (PREP). This process is used for the production of high-reactive powders. Melting of the electrode is carried out in an inert atmosphere. Powder particles produced by rotating electrode processes have a spherical shape with smooth and high-quality surfaces. The particle size distribution is from 50 to 400 μm with D_{50} around 200 μm. In spite of all advantages, there are also disadvantages of these methods. A major of them is a limitation of rotational speed, which restricts the minimum of median particle size to about 50–150 μm. Also, the production of high-quality metal electrode has a high cost; productivity is low; and energy consumption is high compared to other atomizing processes. In the second type of centrifugal atomization, a molten stream of metal is allowed to fall onto a rotating disc or cone, which disperses the melt on droplets under centrifugal forces [9, 10].

2.1.2. Mechanical milling

Mechanical milling was long time one of the most widespread method for the production of iron powder [11]. The conversion of raw material into powder form with mechanical milling occurs in a solid or liquid state. Milling of solids is meant to reduce the primary raw size by destroying them under influence of external forces. There are three types of milling process—crushing, grinding, and attrition. It is possible to combine the different types of treatment of material for reaching the purpose: compression (static), collision (dynamic), shear (incision). The first two types allow to obtain a large size of particles, and the second and third types are used for receiving of fine powders.

Mechanical alloying is a completely solid-state powder processing technique. The process consists of repeated welding, fracturing (crushing), and rewelding of powder particles in a high energy mills. The process due to high intensity of impaction on the fine particles allows to receive powders with non-equilibrium phases (metastable crystalline and quasicrystalline phases), amorphous alloys, nanosized structure. Also the process is used to produce and develop new materials and alloys such as amorphous alloys, intermetallic compound, supersaturated solid phases, and metal matrix composites. The process is used to produce a variety of materials and alloys: supersaturated solid solutions, amorphous materials, intermetallic compounds, and metal-matrix composites [10, 12].

Different types of milling equipment can be used for mechanical alloying, such as horizontal and vertical attritors, disintegrators, planetary ball mills, shaker mills [10].

The advantage of mechanical milling process consists in the possibility to use different raw materials, and it can be pure components, sponge, fibers, or ore for alloying or waste products of mechanical production: chips, shavings, flakes etc.

2.2. Physical–chemical methods

2.2.1. Electrolysis

Electrolysis is known as a physical–chemical process that consists in allocation of electrode components, which occur when the solution or electrolyte melt carry current. Raw material for electrolysis is a metal anode, and in some cases, it is possible to use pressed or sintered waste metal products, choosing needed conditions (composition and viscosity of electrolyte, current density, temperature, etc.) metals can be deposited in powder form. The limitation of electrolysis means an ability to receive pure metal, but not alloys [9].

2.2.2. Chemical processes

The leading of chemical process is a carbonyl process, which allows to produce nickel and iron powder. The crude metal reacts with gaseous carbon oxide under pressure and temperature that lead to the formation of carbonyl, which is decomposed under raising temperature and lowering pressure to metal powder.

Other chemical conversion processes include the following:

– The manufacture of powders from sponges by thermally decomposing chlorides.

– The manufacture of powders by hydrogen reduction of salts solution under pressure.

– Chemical precipitation of metals from solutions of soluble salts. [9]

2.2.3. Plasma spheroidization

One of the new technics for powder production for additive manufacturing is plasma spheroidization. In fact, this is not the method for the production but method of additional treatment of non-spherical powder, which allows to change the shape of particles to ideal spheres.

The world leader in production of plasma spheroidization equipment is a company Tekna. The company's line of products consists of four systems: from laboratory-scale to industrial-scale serial production. Depending on parameters of initial powder, it is possible to make controllable process of full melting and get spherical form of particles during the flight through plasma chamber.

The process benefits do not limited by changing of the shape of particles; it also decreases internal porosity of powder, improves powder flowability, increases apparent density, and enhances powder purity. The last one is quite strong benefit for posttreatment powders after several uses in additive manufacturing [13].

3. Methods of characterization of metal powders properties

In this part of the chapter, standardized methods for the characterization of metal powders will be briefly presented and given that an information about the methods is not standardized, but allow to receive an additional information about properties of metal powders. First, it should be noted that there exists a Technical Committee 119 at the International Standards Organization, since 1967 it has developed and published numerous powder metallurgy standards; most of them have adopted national versions.

All standards of powder metallurgy can be divided in two groups: standards that are similar to material characterization (they are quite typical for non-powder materials) and standards for the characterization of properties of powder.

The first group includes general standards for the determination of common properties of material that relates to typical methods applicable to compact (non-powdered) materials:

determination of chemical composition and determination of interstitial elements. Chemical composition is usually determined by X-ray fluorescence spectrometry, wavelength-dispersive X-ray fluorescence spectrometry, direct current plasma, or inductively coupled plasma atomic emission spectrometry. Determination of interstitial elements is very important especially for reactive metals and alloys, and it is also important to control oxygen content, because the oxygen content may change after several reusing of metal powder due to heat affecting in additive manufacturing. It is very important to pay attention on O, H, and N content in titanium, tantalum, aluminum, and their alloys; on O, C in refractory and reactive metals and their alloys, and steels and nickel alloys.

The second group of standards includes methods for the determination of next properties of powders: particle size distribution; sieve analysis; flowability; apparent density; skeletal density and determination of porosity; and shape of particles.

Particle size distribution is one of the most important properties of metal powder for the application of additive manufacturing. All AM-system producers recommend to use powders prepared and supplied by manufacturer of AM system, and also manufacturer gives recommendations for particle size distribution of powders applicable to their systems. Particle size distribution is usually measured by laser diffraction methods, and a typical report of measurement has a graph and table with values of particle sizes and their volume. The general characteristics are D_{10}, D_{50}, and D_{90}, which mean that volume of 10, 50, and 90% particles has a size smaller than respective values.

Sieve analysis is commonly used to change particle size distribution, for example, to separate huge particles. A typical sieve analysis involves a nested column of sieves with wire mesh cloth (screen). It is possible to make particle size distribution suitable to requirements and recommendation of AM-system manufacturer by sieving.

Flowability of powder effects on smooth coating and equable feeding of powder in AM systems. The main parameters that have an influence on flowability are particle size distribution, density of metal or alloy, shape of particles, and morphology of their surfaces and humidity. Very fine powder (smaller than 10 μm) typically has a poor flowability or do not flow at all, but powder compositions that content fine or big particles have a good flowability. Density of metal or alloy makes effect because the general principle of flowability is to measure the time of flow through funnel (Hall flowmeter) with 2.5 mm diameter orifice 50 g of powder under itself weight, so if the metal or alloy has high density, powder of this metal or alloy will flow faster. Spherical powder flows better than powder with irregular form, because particles do not cling each other. Humidity of powder makes effect of sticking particles together and leads to getting worse results of flowability measurement, so it is strongly recommended to dry powder before using.

Apparent density is the method for the measurement of density of powder compact in a density cup (25 cc) which was received by free flow of powder through a funnel in the density cup. Particle size distribution and shape of particles have influence on apparent density.

Skeletal density shows true solid state density of alloy or powder material. Density depends on quantity of alloying elements, their content in the alloy, and phase composition of material.

The determination of skeletal density is made by pycnometry methods. The physical principle of pycnometry is volumetric displacement by fluid and calculation the ratio of the mass to the volume occupied by that mass.

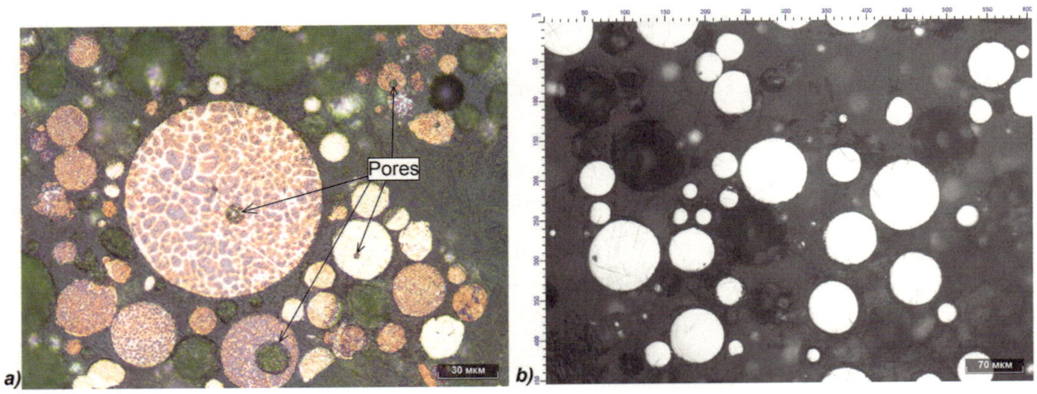

Figure 1. Images of cross section of X22CrMoV powder particles received by gas atomization with internal porosity (a) and Ti–6Al–4V powder received by plasma atomization (b).

It is often used a gas pycnometry where helium or nitrogen is used as fluid medium, because these gases have small atomic sizes and have possibility to penetrate in defects. Sometimes, it is used a liquid pycnometry where dispersion of liquids with high-penetration properties is used as fluid medium (ethanol, oils, butanol, acetone etc.). Measuring of skeletal density is important for the estimation of quantity of defective particles with cracks, satellites, opened and closed pores (see **Figure 1**). Pycnometry also may be used for analyzing of compact materials with irregular shape. For the determination of internal porosity of powder particles, preparation of cross sections and investigation by optical of electron microscopy can also be used.

There is no international standard for measuring of *particles shape*, but there exist national standards (for example, American ASTM E20 and Russian GOST 25849) that content approaches for the description and classification of metal powders by shape (**Figure 2**).

Determination of shape of particles can be made by optical microscopy, but more representative results may be obtained by scanning electron microscopy (SEM). A shape of particle depends on technology, on which a powder has been made. Spherical and spheroidal shape is more specific for the atomization technologies; angular form is typical for mechanical milling and mechanical alloying; dendritic, rod, needle like, and particles with internal void are obtained by electrolysis and chemical processes; plate-like and flaky powder can be produced by mechanical milling in shear mode.

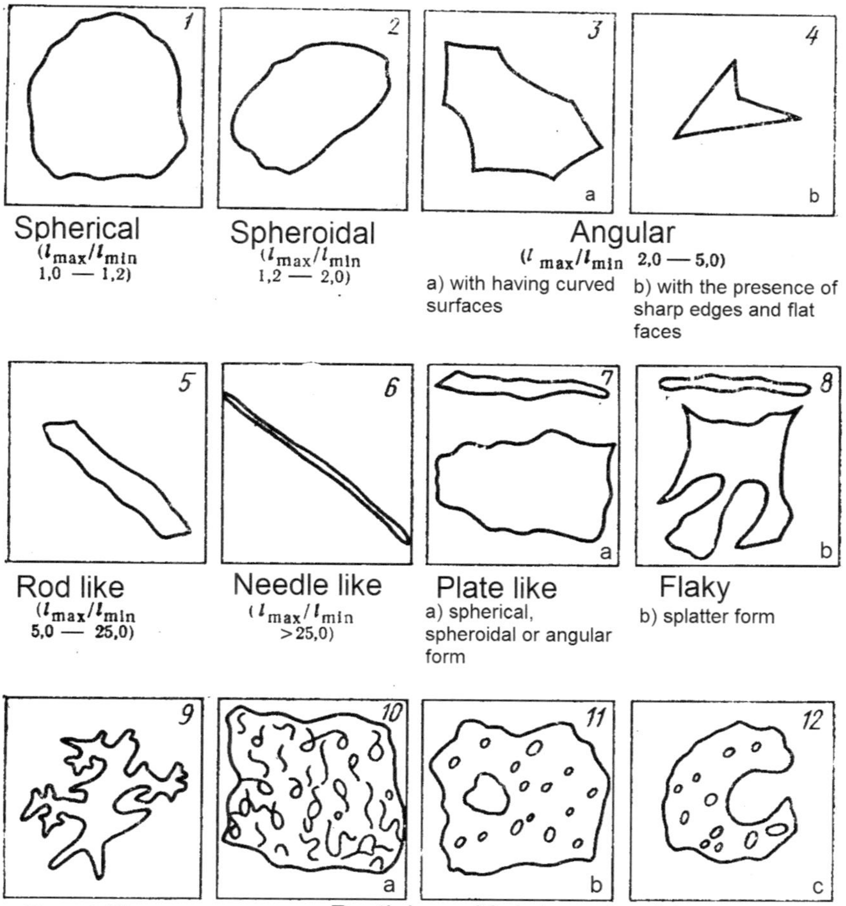

Figure 2. Shapes of powders according to GOST 25849.

In **Figure 3**, it is shown the scanning electron microscopy images of powders with different shapes.

International standards were developed for traditional powder metallurgy technologies of compacting (hot and cold pressing, hot and cold isostatic pressing, metal injection molding etc.), and additive manufacturing technologies have some particularities, so at this moment, an actual purpose consists in developing of methods of determination of properties of metal powders for AM applications.

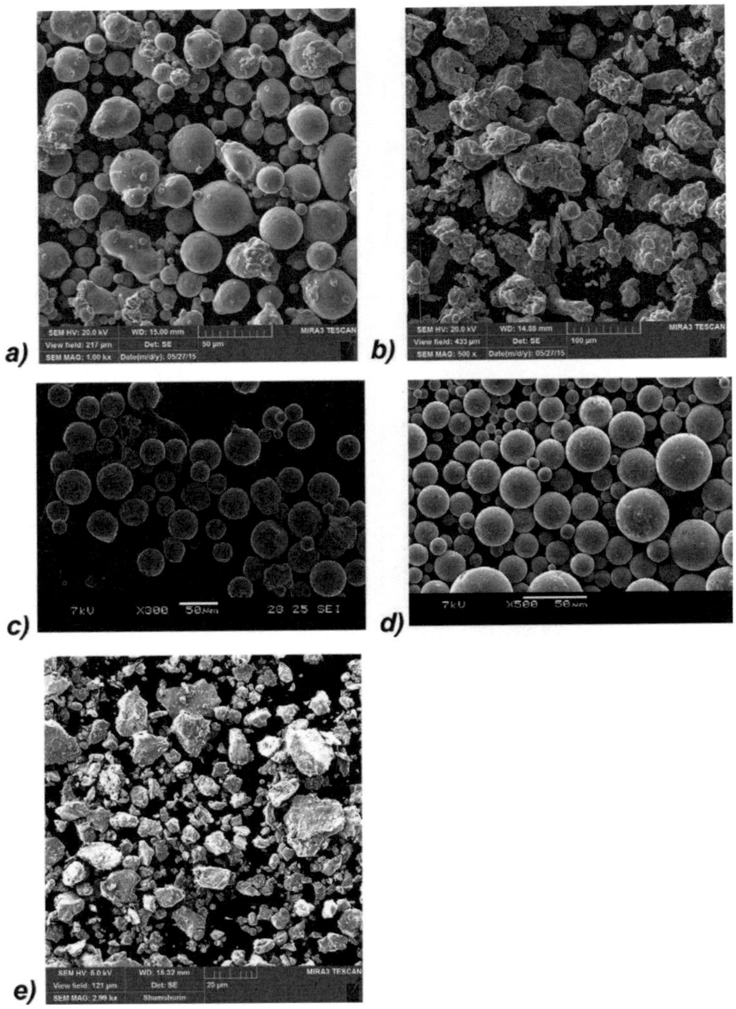

Figure 3. SEM images of powders obtained by different technologies. (a) gas atomized In718; (b) chemical reduction Fe; (c) gas atomized Ti–6Al–4V; (d) plasma atomized Ti–6Al–4V; and (e) mechanically alloyed Fe–18Cr–8Ni–12Mn–N.

One of such methods is described in work [14]. The method is based on measuring of dynamic properties of powder. For measuring, FT4 powder rheometer was used, which allows the measuring of shear, dynamic, and bulk properties. Dynamically determined powder properties are particularly more helpful for defining flowability under the low stress conditions that apply to the most parts of AM process.

Another promising method for testing of powder material was named revolution powder analyzer. The revolution powder analyzer consists of rotating drum covered on the both sides with transparent glass and camera that records pictures of rotating drum (0–200 min^{-1}) before backlight. This method allows the modelling of powder behavior during the coating in powder bed in additive manufacturing systems. As a measuring parameter, the angle of linear regression of the free powder surface measured to a horizontal line is used, just before an avalanche starts [15].

4. Additive manufacturing technologies and properties of parts produced from metal powders

At this moment, there are three main technologies for additive manufacturing from metal powders: powder bed fusion, directed energy deposition, and binder jetting (**Figure 4**).

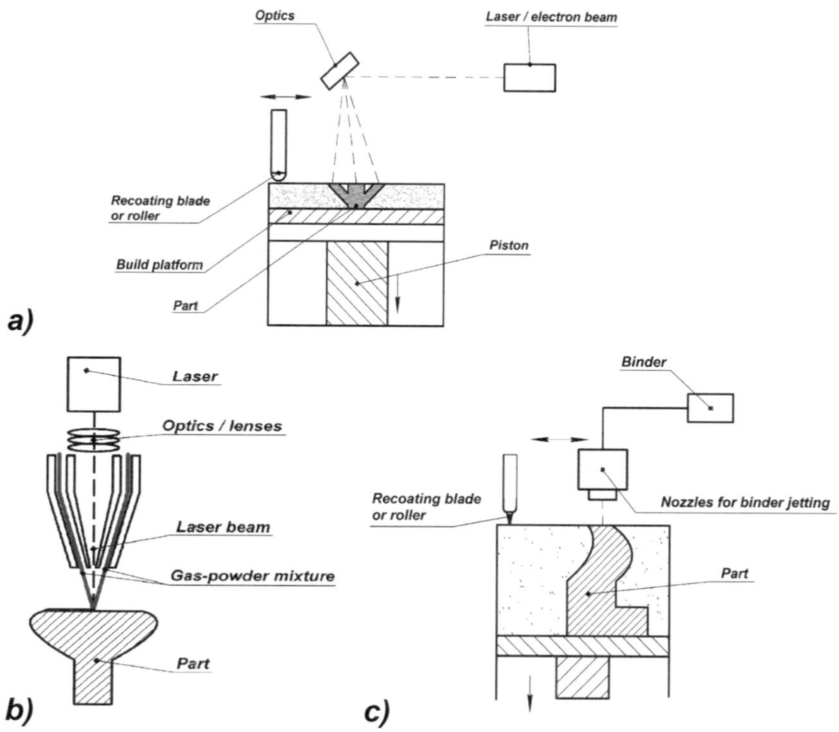

Figure 4. Technological schemes of powder bed fusion (a), directed energy deposition (b) and binder jetting (c).

For powder bed fusion technology, AM-system manufacturers usually use laser as an energy source [EOS, Concept Laser, SLM Solutions, 3D Systems (ex Phenix Systems), Renishaw, Realizer], but there is one company that offers systems with electron beam (Arcam).

The use of electron beam has some features: First of all, electron beam may effectively work only in high vacuum (laser systems work in inert gas atmosphere), and this is a good advantage in working with high-reactive metals and alloys such as titanium; the second one is that before selective melting, whole layer of powder treated by multiple passes of low power electron beam for heating and sintering powder bed, this gives some limitation in geometry, because the sintered powder has to be removed after building.

Laser based on the powder bed fusion systems has differences among themselves. EOS, Concept Laser, and 3D Systems of AM systems feed initial powder from the neighboring to the main build platform tank, whereas SLM Solutions, Renishaw, and Realizer systems feed the initial powder from the main tank which is placed in the upper part of the system. This difference may have an influence on needed properties of powder (flowability). One more difference consists in recoating mechanism, and 3D Systems has patented a mechanism with roller, whereas the others use blades and the use of roller may expand a range of available for the process powders and give an advantage in using of fine powder (less 10 μm) with poor flowability.

Directed energy deposition is usually called cladding. Manufacturers of this type of systems use laser as an energy source, and powder is usually fed coaxial to a laser beam with inert gas. Depending on the cladding nozzle, it is possible to manage speed and accuracy (coaxial nozzle gives the highest accuracy, off axis is the fastest), but anyway, it is impossible to build very complex part such as lattice structures, closed cooling channels with this technology. Advantages of the technology consist in capability to deposit more than one material simultaneously, creating functionally graded coatings and parts. Most directed energy deposition systems use a 4- or 5-axis motion system or a robotic arm to position the deposited head, so the build process is not limited to successive horizontal layers on parallel planes. This capability makes the process suitable for adding of material to an existing part, such as repairing a worn part or tool [1].

Binder jetting is a process, by which a liquid bonding agent is selectively deposited through inkjet print head nozzles to join powder materials in a powder bed. Binder jetting is similar to material jetting in its use of inkjet printing to dispense material. The difference lies in the fact that the dispensed material with binder jetting is not a build material, but rather a liquid one, which is deposited onto a bed of powder to hold the powder in the desired shape [1].

Producing of the parts with binder jetting technology includes 3D printing, debinding, sintering, and sometimes infiltration by another material. The advantage of binder jetting technology consists in lack of need to use support structures; powder bed makes this role; and the absence of high-temperature gradients and phase transformations allow to save desired shape of a future part. Binder jetting in this moment has limited success for producing of metal parts and looks more promising in manufacturing ceramic parts because of its multi-step process, or because the final properties of metal parts are not very high.

The authors have an experience and made researches in the field of using of laser powder bed fusion system. The results of selected researches in this field of additive manufacturing will be presented. Properties of metal part manufactured by selective laser melting (SLM) process (here and further, this name will be used for laser powder bed fusion of additive manufacturing technology) have strong dependence on parameters of process. The main parameters of SLM are layer thickness, laser speed and power, hatch distance, strategy of hatching. Layer thickness is very important parameter, because of its dependence on powder. It is possible to manage speed and accuracy of SLM process by changing of layer thickness; sometimes, it is not very important to have a high accuracy in Z-direction. It gives an opportunity to use larger layer thickness that may increase build speed more than two times. At this moment, most powder bed fusion systems' manufacturers use 400 W laser in their systems, and "standard" layer thicknesses are 20 or 30and 40 or 50 μm. Layer thickness determines maximum of particle size that can be used in process, and particles with sizes more than layer thickness physically will not take part they will be thrown off by recoating blade or roller. Of course, it should be taken in account that there is some changing of density between apparent density of powder after recoating and density of material after melting; additionally, there is shrinkage effect that changes the real layer thickness during the process and depends on type of material, also some volume of particles with large size needed to save flowability properties of powder. That is why it is usually recommended to use a 10–63 μm, 10–45 μm powder in depends on density of alloy (in referring flowability) and layer thickness.

Use of powders with fine particles (for example 0–45 μm) has some ambiguity. The presence of fine particles increases apparent density, which may increase final density after SLM in the same time PBF system manufacturers do not recommend to use powders with fine particles because of the danger of their falling into the working mechanisms of systems. Researchers from University of Nottingham have made investigation about the effect of particle size distribution on processing parameters [16]. They have the following results: Final density after SLM is higher with using of powder 0–45 μm, but strength properties are higher with using of powder 10–45 μm. Another important result consists in the fact that parameters of selective laser melting for reaching the maximum density were different for powder with particle size in range of 0–45 and 10–45 μm. One more research about an influence of powder particle size distribution on properties of final part is presented in [17].

SLM is characterized by rapid laser treatment with melting and solidification of metal, and the process was accompanied with active spark formation. For sparks removing and fuming, it is usually used as the creation of "wind of inert gas" above powder bed which blowing out the sparkles and fume from working zone. In **Figure 5**, it is shown the SEM images of particles that were blown out by "wind of inert gas".

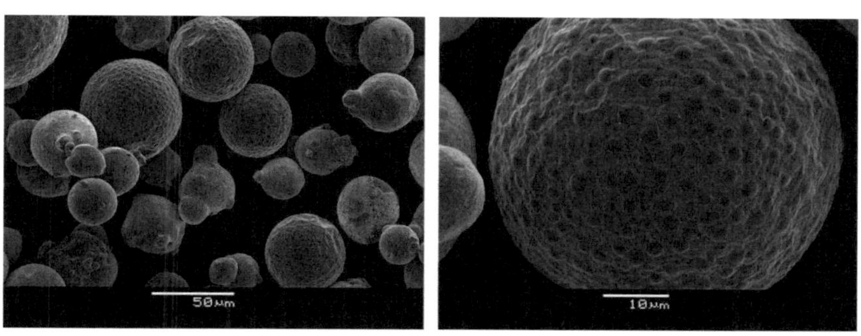

Figure 5. SEM images of In718 fume powder after SLM.

As it seen from the figure/as the figure shows, some particles have dots, which may be some effect of oxidation or changing phase of the composition of powder (In718—gamma prime precipitation hardened nickel superalloys). Typically, the powders received by atomization technologies have a single phase (thanks to rapid solidification during atomization), and this fact makes an applying of powders in AM technologies easier, because the different phases may have different properties (physical density, coefficient of laser absorption, thermal conductivity etc.) and make influence of the process. Powders reuse with some fume content is an actual task for research at this stage of developing of AM.

Another important theme in reusing of powders in SLM process is an agglomeration of particles and loss of spherical shape. Some quantity of particles, lying near to manufactured parts, has been taken by heat effect that leads to sintering with each other. Such agglomerates may have large sizes, and they will be separated by sieving. But there also exist agglomerates from the small particles which can move through sieve (see **Figure 6**).

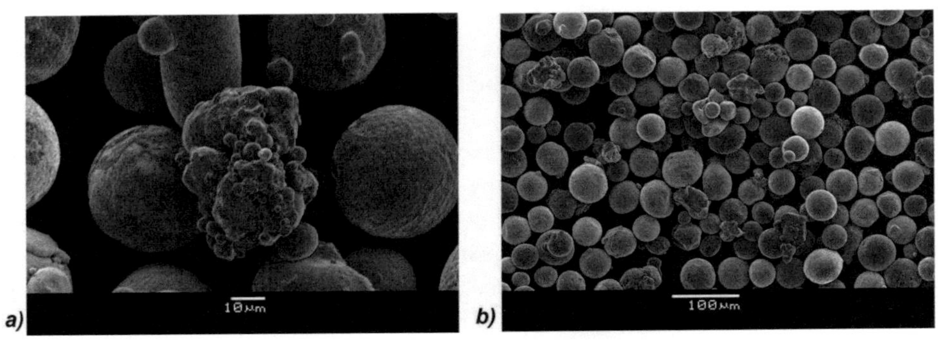

Figure 6. SEM images of In718 (a) and Ti–6Al–4V (b) powders used in SLM.

Phase composition of agglomerates may be different in comparison of virgin powder, during subsequent reusing of quantity of such agglomerates will grow and quality of final parts may decrease.

Mechanical properties of metal parts manufactured by SLM are usually higher than cast metal and sometimes comparable with wrought materials (see **Table 1**) [18–24]

Sample		**Yield strength, MPa**	**Ultimate strength, MPa**	**Elongation δ, %**	**KCV, J/cm²**
Inconel 718 (SLM)	Horizontal	807 ± 15	1051 ± 18	22 ± 4	83.8 ± 3.5
	Vertical	675 ± 12	957 ± 15	28 ± 3	91.3 ± 4.0
Inconel 718 (SLM + heat treatment)	vertical	1160	1350	17.6	
Inconel 718 (casting) [20]	–	477	752	33.8	
Inconel 718	–	1048–1116	1288–1341	21–27	

Sample		Yield strength, MPa	Ultimate strength, MPa	Elongation δ, %	KCV, J/cm²
(Hot-rolled + heat treatment) [21]					
Ti–6Al–4V (SLM)	Horizontal	1200 ± 19	1280 ± 17	2.5 ± 1.0	9.9 ± 2.5
	Vertical	1080 ± 20	1160 ± 16	4.0 ± 1.0	11.7 ± 2.0
Ti–6Al–4V (SLM + heat treatment)	Horizontal	1012 ± 6	1088 ± 6	10.4 ± 1.8	11.9 ± 1.7
	Vertical	945 ± 15	1087 ± 17	10.7 ± 1.5	15.2 ± 2.0
Ti–6Al–4V (wrought + heat treatment) [22]	–	>826	>895	>10	

Table 1. Properties of samples from In718 and Ti–6Al–4V manufactured by SLM and traditional technologies.

There is anisotropy of mechanical properties of samples manufactured parallel (horizontal samples) and perpendicular (vertical samples) according to the main platform of SLM system. The reason of anisotropy is a layer-based synthesis (grain microstructure is elongated in Z-direction). Also flat defects in X–Y plane and residual stresses influence on anisotropy of mechanical properties (**Figure 7a**). Effect of anisotropy may be decreased by heat treatment (stress relief, stable microstructure, and phase composition) and hot isostatic pressing (closing internal defects such as pores and cracks). High residual stresses during SLM is one of the limitation of this technology (see **Figure 7b**), and for solving of this problem, it should be used special strategies of hatching (for example, "chessboard hatching") and carefully prepared support structures.

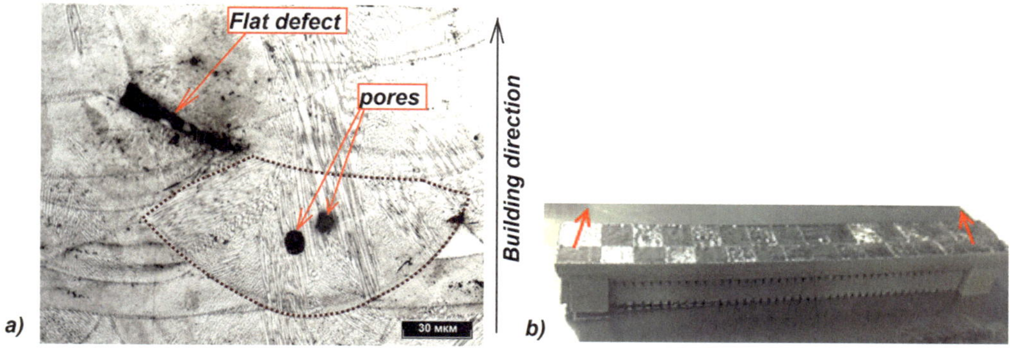

Figure 7. Internal defects after SLM (a) and residual stresses (b) influence on building a sample.

As it was already noticed, selective laser melting is a process with high melting and cooling rates. This fact affects on the microstructure and phase composition of manufactured metal or alloy. It is common to make heat treatment after SLM, and a type of heat treatment strongly depends on a type of alloy; for example, for single phase of austenite stainless steels (such as 316 L), stress relieve annealing might be enough, but for precipitation of hardened nickel superalloy (such as Inconel 718), it needs to make multistage heat treatment (homogenization and aging). The **Table 2** shows the results of XRD analysis of Inconel 718 and Ti–6Al–4V samples [20, 21].

Sample	Qualitative composition	Quantitative composition [vol%]
Inconel 718 powder	γ-Ni	90.0
	γ'-Ni3Al	3.5–3.9
	γ″-Ni3Nb	4.3–4.5
	δ-Ni3Nb	1.8–2.0
Inconel 718 SLM before heat treatment	γ-Ni	86.8
	γ'-Ni3Al	1.9
	γ″-Ni3Nb	8.0
	δ-Ni3Nb	3.3
>Inconel 718 SLM + homogenization	γ-Ni	90.1
	γ'-Ni3(Al,Ti)	1.9
	γ″-Ni3Nb	8.0
Inconel 718 SLM + homogenization + aging	γ-Ni	67.3
	γ'-Ni3 (Al,Ti)	8
	γ″-Ni3Nb	4
	δ-Ni3Nb	3.5
	γ'-Ni3Al	17.2
Ti–6Al–4V powder	α'-phase	100
Ti–6Al–4V SLM without heat treatment	α'-phase	94.49
	β-phase	5.51
Ti–6Al–4V SLM with heat treatment	α-Ti	11.3
	α'-Ti	73.8
	β-Ti	14.9

Table 2. The results of XRD analysis of Inconel 718 and Ti–6Al–4V samples.

Synthesis of initial powder material is a result of high-speed solidification of the melt droplets in an inert gas stream, that is, crystallization takes place under non-equilibrium conditions, which affects the completeness of the phase transition.

In Table 2, it is shown the changing of powder phase composition and compact samples after SLM and heat treatments. The powder of Inconel 718 and the compact sample after SLM have a similar high content of γ-Ni matrix-phase which is a result of rapid solidification, but due to the presence of heat-affected zones, the quantity of γ″-Ni3Nb and δ-Ni3Nb phases in the compact sample is higher. Heat treatments of the compact samples lead to the changing of

http://dx.doi.org/10.5772/63337

phase composition: Homogenization dissolves δ-Ni3Nb phase, aging increases quantity of precipitates.

The study of the phase composition of the initial powder alloy Ti–6Al–4V showed that the powder consists of more than 99% of α'-phase. Qualitative phase composition of the compact sample after SLM is different from powder material by the presence of β-phase (its content is 5.51%).

Changing of the phase content also can be seen on the microstructure investigations (**Figures 8** and **9**).

In Figure 8, microstructure of In718 samples after SLM, homogenization, and aging is presented. Grains of γ-Ni are elongated along building direction (Z-axis) and have different size from 10 to 200 μm. Coagulated precipitates are uniformly distributed and have size of 4–5 μm. Some precipitates lined up in chains with length up to 10 μm; however, there is not seen full edging of γ-Ni grains. Also some of precipitates observed inside of γ-Ni grains.

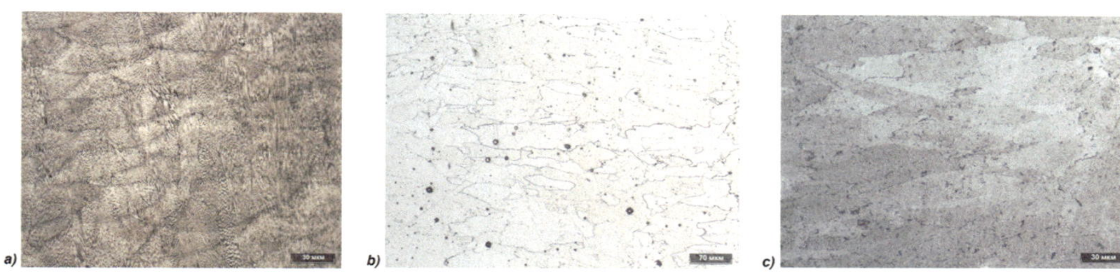

Figure 8. Microstructure of cross section of Inconel 718 specimens, manufactured by SLM, before heat treatment (a), after homogenization (b) and aging (c).

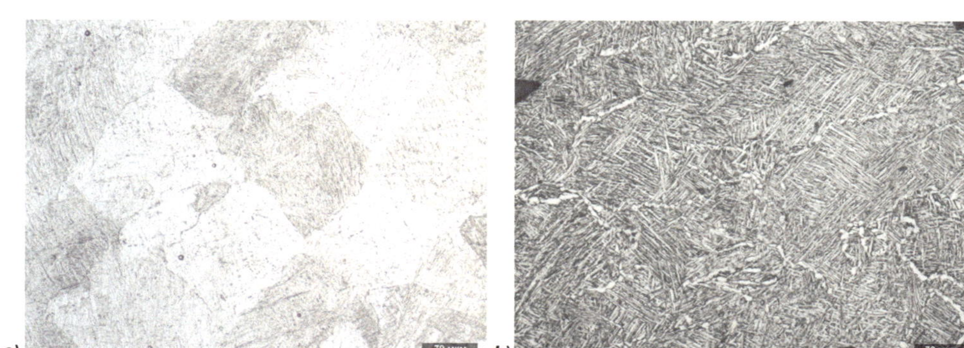

Figure 9. Microstructure of cross section of Ti–6Al–4V specimens, manufactured by SLM, before heat treatment (a) and after annealing (b).

Before heat treatment, Ti–6Al–4V specimen has a basket type of microstructure (see **Figure 9**). After annealing, some α- and β-phases stood at grain boundaries, initial martensite of needles enlarged in size, borders become more rounded compared to the sample without annealing.

The study of fractography of fracture surfaces of impact strength specimens showed some imperfections of SLM process (see **Figure 10**).

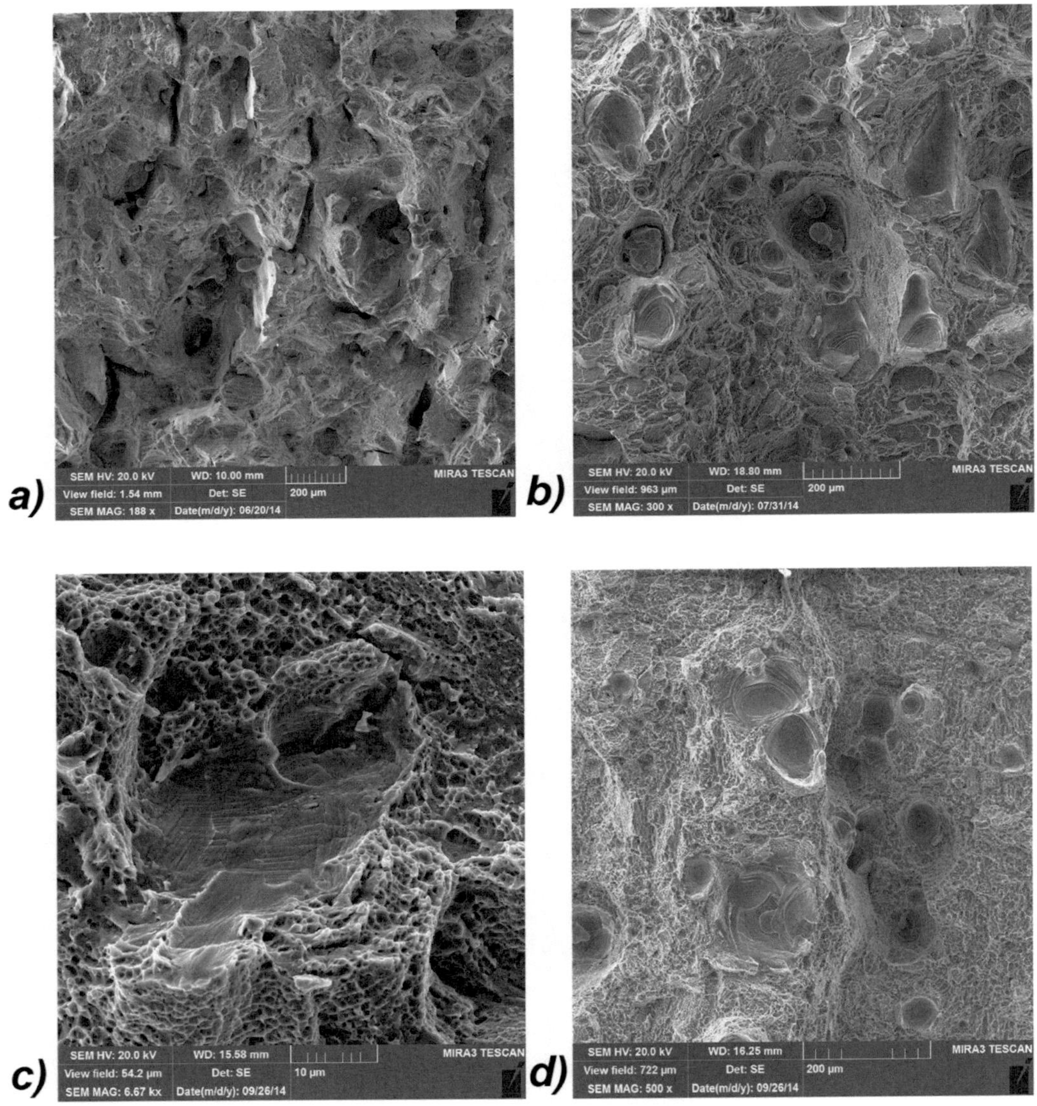

Figure 10. Fractography of Inconel 718 (a, c) and Ti–6Al–4V (b, d) specimen fracture surfaces before (a, b) and after heat treatment (c, d).

As shown in **Figure 10**, there are some micropores on the fracture surface, which function is stress concentration for cracks growth. Some micropores contain not melted powder particles.

5. Conclusion

In this chapter, a review of powder production methods, characterization of metal powder with focus on the application and technologies in additive manufacture, which use metal powders as initial material, was presented. The researches in the field of microstructure and properties of samples, which are produced by selective laser melting, also were presented.

The main technologies for mass production of metal powder with spherical or spheroidal particle shape relate to the atomization methods (dispersion of metal melt). However, the other technologies may cost cheaper because of using the waste products (chips, shaving, flakes etc.) as raw materials and the approach of receiving of powders with subsequent plasma spheroidization looks very promising, especially for developing of new alloys that impossible to produce by melting technologies.

Some standardized methods for the characterization of metal powders help to understand differences between powders produced by different technologies, and also they may be used for fixation the evolution of powder properties after reusing in additive manufacturing. Developing of special methods for checking the properties of powder gives an additional information about behavior of powder in AM systems, may significantly expand a range of applicable powder in AM, and allows to better understand requirements for powder for using them in additive manufacturing.

There exist three main technologies for additive manufacturing from metal powders: powder bed fusion, directed energy deposition, and binder jetting. Every of them has advantages and disadvantages, and at this moment, they all are in intensive developing phase all around the world. The main advantage of all types of additive manufacturing is a possibility to produce parts with design that impossible to manufacture by traditional technologies. It is necessary to conduct huge researches and tests to be sure that produced by AM parts have reliable and repeatable properties.

6. Outlook

At the moment, the list of metal alloys, available for additive manufacturing, is not long. Research and developing of applying of metal powders received by different technologies will expand the list and make additive manufacturing more accessible and economical for different areas of application. Unique possibilities of control of energy source tracing and using of different scanning strategies in each layer and make additive technologies very complex, but at the same time, it is quite promising for future developing. Application of complex geometry of parts with lattice structures, cells, and others, produced by additive manufacturing, is not fully revealed and will amaze not once by area of application in future.

Acknowledgements

The authors are grateful to Evgernii Borisov, Igor Polozov, Dmitry Masailo, and Ivan Goncharov for help in researches and preparing an information for this chapter. Also, the authors want to thank Maxim Maximov and Aleksey Shamshurin for researches with scanning of electron microscopy.

Author details

Anatoliy Popovich and Vadim Sufiiarov*

*Address all correspondence to: vadim.spbstu@yandex.ru

Peter the Great Saint–Petersburg Polytechnic University, Polytechnicheskaya, St. Petersburg, Russia

References

[1] Wohler, T., editor. Wohler report. 19th ed. Wohlers Associates, Inc; 2014. OakRidge Business Park, 1511 River Oak Drive Fort Collins, Colorado 80525 USA. 277 p.

[2] SLM Solutions GmbH. Available from: http://slm-solutions.com [Accessed: 10.02.2016].

[3] ExOne. Available from: http://www.exone.com/ [Accessed: 10.02.2016].

[4] BE Additive Manufacturing. Be Additive Manufacturing [Internet]. Available from: http://beam-machines.fr/ [Accessed: 10.02.2016].

[5] Fabrisonic. Ultrasonic Additive Manufacturing [Internet]. Available from: http://fabrisonic.com/ultrasonic-additive-manufacturing-overview/ [Accessed: 10.02.2016].

[6] Sciaky Inc. Electron Beam Additive Manufacturing [Internet]. Available from: http://www.sciaky.com/additive-manufacturing/electron-beam-additive-manufacturing-technology [Accessed: 10.02.2016].

[7] ALD Vacuum Technologies. Metal Powder Production Equipment [Internet]. Available from: http://www.ald-vt.com/cms/fileadmin/pdf/prospekte/Metal_powder.pdf [Accessed: 10.02.2016].

[8] AP&C. Plasma Atomization [Internet]. Available from: http://advancedpowders.com/wp-content/uploads/2014/02/Raymor-APC-Leading-the-way-with-plasma-atomised-Ti-spherical-powders-for-MIM.pdf [Accessed: 10.02.2016].

[9] Powder Metallurgy Review. Powder Production Technologies [Internet]. Available from: http://www.ipmd.net/Introduction_to_powder_metallurgy/Powder_Production_Technologies#sthash.qNceqYyC.dpuf [Accessed: 10.02.2016].

[10] Neikov, O.D., Murashova, I.B., Yefimov, N.A., Naboychenko, S. Handbook of non-ferrous metal powders: technologies and applications. Amsterdam: Elsevier; 2009. 616 p.

[11] European Powder Metallurgy Organization. Powder Metallurgy Manufacturing [Internet]. Available from: http://www.epma.com/powder-metallurgy-powder-manufacture [Accessed: 10.02.2016].

[12] Popovich, A., Razumov, N. Dissolution of alloying elements and phase formation in powder materials Fe–18Cr–8Ni–12Mn–xN during mechanical alloying. Advanced Materials Letters. 2014;5(12):683–687. doi:10.5185/amlett.2014.6585

[13] Tekna. Spheroidization Equipment [Internet]. Available from: http://tekna.com/equipment-spheroidization-nanosynthesis-deposition/spheroidization-equipment/ [Accessed: 10.02.2016].

[14] Clayton, J. Optimising metal powders for additive manufacturing. Metal Powder Report. 2014;69(5):14–17. doi:10.1016/S0026-0657(14)70223-1

[15] Spierings, A.B., Voegtlin, M., Bauer, T., Wegener, K. Powder flowability characterisation methodology for powder-bed-based metal additive manufacturing. Progress in Additive Manufacturing. 2015 (http://link.springer.com/journal/volumesAndIssues/40964);1–12. doi:10.1007/s40964-015-0001-4

[16] Liu, B., et al. Investigation the effect of particle size distribution on processing parameters optimisation in selective laser melting process. In: International solid freeform fabrication symposium: an additive manufacturing conference. Austin: University of Texas at Austin; 2011. pp. 227–238.

[17] Averyanova, M., Bertrand, P.H., Verquin, B. Studying the influence of initial powder characteristics on the properties of final parts manufactured by the selective laser melting technology: a detailed study on the influence of the initial properties of various martensitic stainless steel powders on the final microstructures and mechanical properties of parts manufactured using an optimized SLM process is reported in this paper. Virtual and Physical Prototyping. 2011;6(4):215–223. doi: 10.1080/17452759.2011.594645

[18] Sufiiarov, V., Popovich, A., Borisov, E., Polozov, I. Selective laser melting of Inconel 718 Nickel superalloy. Applied Mechanics and Materials. 2015;698:333–338. doi: 10.4028/www.scientific.net/AMM.698.333

[19] Sufiiarov, V.S., Popovich, A.A., Borisov, E.V, Polozov, I.A. Selective laser melting of heat-resistant Ni-based alloy. Non-Ferrous Metals. 2015;38(1):32–35. doi:10.17580/nfm.2015.01.08

[20] Sufiiarov, V.S., Popovich, A.A., Borisov, E.V., Polozov, I.A. Microstructure and mechanical properties of Inconel 718 produced by SLM and subsequent heat treatment. Key Engineering Materials. 2015;651–653:665–670. doi:10.4028/www.scientific.net/KEM.651-653.665

[21] Sufiiarov, V.S, Popovich, A.A., Borisov, E.V., Polozov, I.A. Microstructure and mechanical properties of Ti–6Al–4V manufactured by SLM. Key Engineering Materials. 2015;651–653:677–682. doi:10.4028/www.scientific.net/KEM.651-653.677

[22] Sufiiarov, V.S., Popovich, A.A., Borisov, E.V., Polozov, I.A., Maximov, M.Y. Studying of microstructure and properties of selective laser melted titanium-based. Advanced Materials Research. 2015;1120–1121:1269–1275. doi:10.4028/www.scientific.net/AMR.1120-1121.1269

[23] Sufiiarov, V.S., Popovich, A.A., Borisov, E.V., Polozov, I.A. Selective laser melting of Ti–6Al–4V for gas turbine components manufacturing. Non-Ferrous Metals. 2015;39(2): 21–24. doi:10.17580/nfm.2015.02.04

[24] Sufiiarov, V.S., Popovich, A.A., Borisov, E.V., Polozov, I.A. Layer thickness influence on the Inconel 718 alloy microstructure and properties under selective laser melting. Tsvetnye Metally. 2016;(1):81–86. doi:10.17580/tsm.2016.01.14

[25] El-Bagoury, N., et al. Influence of heat treatment on the distribution of Ni2Nb and microsegregation in cast Inconel 718 alloy. Materials Transactions. 2005;11(11):2478–2483.0

[26] Special Metals. Inconel 718 Datasheet [Internet]. Available from: http://www.special-metalswiggin.co.uk/pdfs/products/INCONEL alloy 718.pdf [Accessed: 10.02.2016].

[27] Bibusmetals. Sheets, Plates, Strips and Bars from Titanium Grade 5 [Internet]. Available from: http://www.bibusmetals.com.ua/fileadmin/materials/PDF/catalogs_new_2013/titan/Titan_Grade_5_RU_EN.pdf [Accessed: 10.02.2016].

Chapter 11

Laser-Assisted 3D Printing of Functional Graded Structures from Polymer Covered Nanocomposites: A Self-Review

Igor Volyanskii and Igor V. Shishkovsky

Additional information is available at the end of the chapter

http://dx.doi.org/10.5772/63565

Abstract

As a method for conservation of nanoparticles with perspective properties, the three-dimensional (3D) printing is a promising technique for modeling, fabricating of functional graded structures (FGS) with nanoadditives and functional devices. The stabilization of nanoparticles in a polymeric matrix and additionally reinforced porous structure makes it possible to arrange a desired distribution of the nanoparticles in the polymer and thus to protect them from oxidation and corrosion and even to design not only the FGS but also micro/nanoelectromechanical systems (M/NEMS) devices. The synthesized nanocomposites with controlled porosity and large-specific surface may also find their application in implantation, catalysis, lab-on-chips, drug delivery systems, and 3D crystalline structures for hydrogen storage devices.

Keywords: 3D printing, functional nanoparticles, selective laser sintering/melting (SLS/M), functional graded structures (FGS), micro/nano – electromechanical systems (M/NEMS)

1. Introduction

The additive technologies (ATs) (three-dimensional (3D) printing, selective laser sintering/melting (SLS/M), etc.) are promising techniques for modeling, fabricating of functional graded structures (FGS) with nanoadditives and functional devices, but a direct SLS/M fabrication of the nanopowders by multilayered techniques is a difficult technological task. Laser sintering and melting are known to be thermally activated processes accompanied by the coagulation of nanoparticles into micro-sized conglomerates. However, a real challenge is the aggregation prevention during the 3D printing, which significantly levels the potential advantages of using the materials in the nano- or submicron state. One of the ways to solve this problem is the isolation of nanoparticles into the inert matrices where they do not undergo any aggregation or "aging" and can be controllably released with the retained of chemical and phase composition [1–3].

Stabilization of nanoparticles in a polymeric matrix and additionally reinforced porous structure makes it possible to arrange a desired distribution of the nanoparticles in the polymer and thus to protect them from agglomeration, oxidation, and corrosion and even to design the FGS. The results indicate that nanoparticles mechanically reinforced the polymer matrix and elastic modulus and the maximum stress significantly increased. Finally, the correlations "prehistory of obtaining (i.e., "background") - chemical composition of the nanoparticles volume and surface condition - phase/structural composition - morphology - perspective properties" will determine the nanoparticles behavior in further applications.

The present review will demonstrate how laser-assisted techniques of the 3D synthesis could be used to prepare a porous core-shell polymer structures containing different encapsulated nanoparticles distributed heterogeneously over the sintered polymer and dangerous for cancer tissue account of thermal hyperthermia or cytotoxic effect. We demonstrated a principal feasibility for fabrication of functionally graded 3D parts with the structural ordering of iron oxide particles and determined corresponding laser optimal regimes. The SLS-fabricated 3D samples of biocompatible iron oxide core/(polyetheretherketone (PEEK) or polycaprolactone (PCL)) shell magnetic nanocomposites have potential medical application for the tissue engineering scaffolds and cell targeting systems [4].

Functionally graded 3D parts with alternating ferromagnetic Ni-PC and nonmagnetic Cu-PC layers [5] exhibited hysteresis phenomena that can probably be used in microelectromechanical systems (MEMS)-nanoelectromechanical systems (NEMS) applications [6] also, where the time response must depend on the relaxation rate. The synthesized nanocomposites with high porosity and large-specific surface could also find their application in catalysis, lab-on-chips, drug delivery systems, and 3D crystalline structures for hydrogen storage devices [7].

2. Inkjet 3D printing of optical nanostructures

The inkjet 3D printing has already been used for more than 20 years. Up to now, this method remains to be one of the most popular due to its low price both for home using, and for commercial applications. During this period, the techniques of the inkjet 3D printing made an incredible progress [8]. The step of drop resolution advanced from 500 μm (30 dpi) to several micrometers (9000 dpi), reaching the value boundary to the nanoscale range. The absence of high temperatures and a wide variety of inks made a revolution in the field of biosensor and print electronic devices. This method was used for printing first organic light-emitting diode (OLED) displays and panels.

The possibility of the point deposition of thin layer coatings is often used to create conductive layers [9]. Under an intensive study is the possibility of applying the inkjet 3D printing for the graphene electronics fabrication, production of ordered arrays with metallic conductivity and high transparency, for deposition of electrodes of a predetermined shape to a substrate, and also for generation of 2D microarrays in printing electronics. A well-known method is printing of metal precursors (the most popular are Au^{+}, Ag^{+}, Ni^{2+}, Cu^{2+}), followed by the deposition of a reducing agent from another cartridge. These approaches are widely used in industrial printing of electrochemical sensors and combined differential thermocouples, as well as in deposition of substrate electrodes on the conductive polymeric panels.

No less revolutionary is the application of the inkjet 3D printing in the bioengineering field. For example, for the first time the selective bacterial test systems for the analysis of drinking water quality were researched and developed, mechanisms of diagnosing pancreatitis were studied with the aid of biosensors applied to the inkjet 3D printing substrate. Just that very method was used to obtain noninvasive high-sensitive biochips estimating the content of hydrogen peroxide and glucose, as well as to design colorimetric sensors aimed to detect neurotoxins and pesticides [5, 10]. However, the use of the inkjet 3D printing mechanism for bioengineering basically reduced to finding of new approaches to the biomolecules fixation into porous matrices that ensure a stable trapped substance for a long period of time after its deposition on the substrate and drying. Thus, the fabrication of highly inert colloids capable of biomolecules capturing is the most important task for the 3D bioprinting.

The optical nanostructures designed to control the photons flow are of a high practical and fundamental importance. Studies on the interaction of electromagnetic radiation (EMR) with low-dimensional semiconductor and dielectric media suggest the occurrence of a whole set of multifactor processes that jointly determine the probability of photon transport as a key mechanism for the quantum communication. Hence, the control of light flows in nanostructure constructions brings closer to the developing of new photonic devices. Thus, the development of widely available 3D inkjet printing technology for getting optical nanostructures and construction applicable in quantum communications is a pressing problem. This method proved to be successful in the field of electronics and biosensor engineering, including the generation of microchips and microintegrated circuits. The use of the nano-oxide-based system ensures a high stability and repeatability of 3D process. Earlier, we have shown the crystalline phase obtained from liquid solutions with the following condensation of boehmite, anatase, and magnetite under temperatures not exceeding 100°C and without annealing of the samples obtained [11].

Hence, it is possible to fabricate photon-induced panels with a uniform spreading of the light wave in the thin layer. These objects are unique both from the standpoints of the effect proper, and from the viewpoint of their fabrication by 3D methods of a soft chemistry. Such heterostructures will allow to carry out the uniform photon transport (for the specified wavelength) by the whole area perimeter and irrespective of the excitation point. The substrate structure can serve the base for transfer and reading of information in the new PC generation operating by the photon-signal principle. The deposition of planar waveguides on flexible polymeric substrates by the inkjet methods from solutions will greatly simplify and accelerate machine-readable signals pickup on carriers. However, the deposition of transparent dielectric structures with the accuracy of several nanometers by the 3D inkjet technology is still inaccessible.

A complex "structure—property" correlation is required to be determined, ensuring the controllability of optical characteristics of the sample being formed at the stage of the "ink" preparation for the jet printing.

The question of the materials and 3D inkjet technology use for optics remains open. This is due to the fact that the droplet coating aimed to form a solid phase on the substrate surface can embrace a predetermined range of 10^{-3}–10^{-6} m with a high accuracy, while the transition to nanoscale remains uncontrollable. But this nanorange is of a particular importance for the optical use, since nanostructures are commensurable with the wavelength of light in the visible range (400–700 nm).

On the other hand, printing of photonic materials and optical structure is still gaining in popularity, and is solely determined by the development of printing technologies and by increase in their accuracy. The size effect has a major impact on the photon flow control thus imposing many restrictions to the used ink. Still unsolved is the problem of a high heterogeneity of morphology in the drying process (coffee-ring effect). The maximal achievements were obtained under the controlled application of photonic crystals. However, in this case, the light control does not depend on the 3D inkjet technology and is exclusively resulted by the dimensions of spherical particles composing the ink.

As the universal method to control the photon flow, the use of layers with a high refractive index (RI), encased into a core with a low RI can serve. Classically, these approaches are often used for the creation of antireflective coatings and planar waveguides. However, the formation of the high RI transparent layers (more than 2.0 in the visible range) it is not an easy task. To solve this task, attempts were made to modify the polymers by using various nanoscale crystalline additives improving the optical properties of the polymers. In addition to this, the adjustment of 3D printing parameters implies the evaluation of viscosity and surface tension, as well as fine-tuning of the printer for a specific composition. Generally, in order to change the rheology, the additives increasing viscosity, such as glycerin, and surfactants decreasing the surface tension are used [9, 12].

Now only a few of inorganic materials among a great variety of those adapted to the 3D inkjet printing and widely used, can be attributed to highly refractive ones, having high transparency and not expensive, they are for example, ZrO_2, TiO_2, ZnO, and several mixed oxides. TiO_2 and magnetite are considered to be the most versatile for the obtaining of ink with high RI. This soft chemistry approach is well presented in publications, due to known values of the RI for TiO_2 in the anatase phase (2.61), and for Fe_3O_4 (3.2). At the same time, TiO_2 is completely transparent in the visible range, and its colloids are readily formed in an aqueous medium, and Fe_3O_4-based films are magnetically controlled. Printed aluminum and silicon oxides can be used in the polymer environment with a low RI. Now they are the most common matrixes for the storage and delivery of biomolecules. The alumina is used in microelectronics for deposition of dielectric layers, including the use of the 3D inkjet technology. Due to the tendency of these two systems to spontaneous and uncontrolled polycondensation it is necessary to study the influence of sonication on the particles stabilization. For planar waveguides obtaining by the inkjet technology, the multipass printing should be applied. For this, the ink combinations with different RI from cartridges of different types should be alternated. This approach concept consists in using of a layered heterostructure wherein the layer with the minimum RI is formed of spherical nanoparticles with a narrow distribution by size. If the sphere diameter is commensurable with the wavelength of the external light excitation, then the effect of the critical angle of total reflection will be minimized.

3. 3D-printing of polymer nanocomposites, characterized by a tailored viscosity and biodegradability

Nowadays there is no polymer processing treatment capable of competing with the SLS with respect to the fabrication flexibility and complexity of the 3D shapes obtained. The priority of such processing for the construction of new complex parts is clear. However, the advancement of polymer laser sintering technologies is obviously hindered by several factors, such as:

1. Insufficient repeatability and control of the process, which stems from a lack of fundamental understanding. Indeed, the intricate relations between the polymer properties and processing conditions on the one hand, and the final microstructure of the material and/or physical properties of the object on the other hand, are far from being completely understood.

   Polymer powdered materials are characterized by complex physical mechanisms involved in the heating processes by a laser radiation source followed by their coalescence and melting, and then by crystallization during cooling, therefore they require the advanced experimental techniques allowing to study them in real time and at the adequate space resolution [13, 14].

2. Difficulties in developing of polymer matrices with optimized parameters of the melting transition and melt viscosity.

   Since for pure polymer matrices the fine-tuning of these parameters can hardly be achieved, the use of polymer with nanocomposite inclusions is very promissory.

The above-mentioned conclusions are based on the recent studies of the nanocomposite rheology [12–14]. Earlier, in most studies on the polymer nanocomposites, the filler particles of 10–20 nm and even bigger were used. But of highest interest is the case when the size of the particles is still lower than 10 nm. In this case, the particles size becomes commensurable with the polymer ball dimensions, individual polymer coils are approached and the interparticle half-gap is comparable with their gyration radius (R_g) or even smaller than this [15]. This produces chain confinements and distortion that can result in the depletion-driven phase segregation, or unusual properties—if segregation can be avoided. Thus, it is well known that the melt viscosity of polymers can be significantly reduced by adding a small amount of fine nanoparticles. This phenomenon is explained by the increase in the free volume owing to adding nanoparticles that is confirmed in some cases by the decrease in the glass-transition temperature. It should be noted that the viscosity reduction was observed only in cases when the interparticles half-gap was smaller than the R_g of the polymer coil.

The self-healing effect is another interesting example observed in a multilayered nanocomposite polymer structure [16]. The study showed that nanoparticles dispersed in a polymer matrix migrated to cracks generated at the interface between the polymer and glass fiber layer. According to the results of computer simulations, nanoparticles in a polymer can segregate to the surfaces and into the cracks due to the polymer-induced "depletion attraction" between the particles and the surface. In this case, only the particles comparable by their size to the polymer R_g were driven from the matrix to the surface in the crack area. Importantly, a homogeneous dispersion of the nanoparticles in the polymer matrix is a prerequisite for achieving the above-mentioned properties.

The surface enrichment with nanoparticles may result in the improvement of several practically important properties (i.e., friction, wear resistance, flame retardation, chemical resistance, touch feeling, and optical properties). This phenomenon was not taken into account in most studies of the polymer nanocomposites. In recent years, polymer composites are used increasingly for tribological applications, and the effectiveness in tribological performance of nanocomposites over microcomposites has been verified in many systems [17]. However, the specific role of nanoparticles in this case remains an open question. In particular, the tribological properties have rarely been related to the nanocomposite morphology, i.e., the distribution of particles in the polymer (bulk vs. surface). Yet, the improvements of properties can be attributed in this case to the surface enrichment with the nanoparticles.

The selection of new bioresorbable nanocomposites is feasible by the SLS approach. The polymer materials became widely used in various biomedical applications such as design of synthetic tissues and organs from biocompatible materials and stem cells, testing of new drug delivery systems, researches on tissues and organs in normal and abnormal states. The ability to form controllable regular structures from polymer molecules is used for the development of biomedical materials such as porous matrixes for tissue engineering and therapeutic agents' delivery [18]. The fabrication of polymer matrices from biodegradable polymers can facilitate the solution of environmental issues related to the 3D printing industry and promote its further development.

In spite of a great amount of works devoted to the development of polymer nanocomposites, including biocompatible and/or bioresorbable polymers, a clear insights into their structure—property relations is still extremely deficient. This is partly due to the tendency of nanoparticles toward the agglomeration inside the polymer matrix. The fabrication of polymer-based matrix nanocomposites with a homogeneous distribution of particles sized from tens to hundreds of nanometers, affords to systematically study the influence of the ratio of interparticles distance to the polymer coil size on the melt rheology and performance properties of the nanocomposites.

Nanoparticles of different sizes are to be surface-modified and mixed with polymer, so that to ensure a homogeneous particles distribution. Earlier, we received mesocomposites based on PA/PC etc. containing different amounts of inert inclusions (including those of nanosizes) [3, 6]. The use of the nanoparticles is expected to enhance the control over the melt viscosity, crystallization kinetics (e.g., nucleation effect of the nanoparticles), and preferable formation of certain polymorphic modifications of PA/PC (e.g., γ-phase vs. α-phase), which is highly important for SLS processes [13]. The use of ultrasmall nanoparticles for obtaining control over the nanocomposite melt viscosity is original and is based on recent researches of the melt viscosity decrease due to the addition of such nanoparticles to the polymer matrix.

Consideration of the polymer matrices based on biodegradable polymers such as polycaprolactone (PLC) is another interesting issue for the practical medicine [19, 20]. This last issue is of high importance for all the 3D printing technologies, as it can solve the problem of growing concerns related to the environmental threat of the 3D printing industry development. We were among those first who experimentally studied the optical characteristics of polymer systems (PC, PA, etc.), impregnated with metallic particles [21]. We used an advanced strategy of determination the beam part, which was subjected to absorbing, scattering, and transmission. We offered original decisions for the characterization of thermophysical properties of such metal—polymer powdered mixtures also [22]. However, there is an obvious need to continue this work with nano-sized additives.

While studying the heat transfer and phase transformation processes, it is highly demanded to analyze the microstructural model of semicrystalline polymer, which is usually forming the nano-sized crystals (lamels) that have a very small depth but a significantly larger lateral size. The polymer structure formation under laser sintering is described in papers [23, 24]. However, for the comprehensive theoretical description of the process, the knowledge of a great number of parameters is demanded, such as temperature of phase transformation, enthalpy changing and thermochemical properties of materials. Moreover, rheological parameters of the polymer melt will influence the sintering process, which in detail described by the molecular dynamic theory [25]. Besides, the particles aggregation depends on viscoelastic properties of the polymer matrix, which are, in turn, crucially dependent on the environment conditions, as well as on degree of crystallinity and crystal texture of the material. The modeling of the melted particles aggregation behavior can be conducted by using simplified models of Frenkel, Maxwell, Kelvin-Voigt, or Bingham.

Therefore, the study of the nanoparticle influence on the final material properties gains special importance. Oxides nanoparticles are most widely used nano-sized additives for polymers and their synthesis and surface modification are well known. Nevertheless, nanoparticles aggregation is still hardly avoidable. It is known that presence of silicon dioxide nanoparticles in the polymer matrix increases the Young modulus but decreases its ultimate elongation, as to the data on the nanoparticles influence on others properties, the data about it is contradictory and inconsistent.

4. FG nanostructures with different connectivity types, obtained by hybrid SLS technology

The search and synthesis of new electro- and magneto-active materials are almost exhausted due to the almost complete use of the existing chemical compositions and processes of their preparation in various solid states, and also due to the limited functions of samples. Therefore, it is an urgent problem of today to switch over to the multicomponent mesoscopically inhomogeneous (i.e., functional graded, FG) nanostructures with different types of orderings (and different thermal, magnetic, piezoelectric, ferroelastic properties, etc.) [25].

The most promising materials are lead-free phases (involving niobates of alkali metals and alkaline earth metals, bismuth ferrite, etc. with various additives) [26, 27]. These materials have giant macroresponses including ultrahigh Curie ($T_C \geq 1400$ K) and Neel ($T_N \geq 1000$ K) temperatures providing a wide range of practical applications. They are ferroelectric, antiferroelectric, piezo- (magneto) electric solid solutions, and/or their based compositions due to greater polyfunctional characteristics in vicinity of new structured phases appearance with their accompanying extreme electro-(magneto-) physical parameters.

However, there are some negative factors impeding their application, which are related to the physical and chemical features of these objects (decomposition during the process of heat treatment, problems with poling, high volatility of the starting components, excessive grain growth due to recrystallization, low thermal stability, and mechanical strength, etc.), which could be solved by additive technologies. An urgent and significant problem of modern physical material science is the development of experimental and theoretical base of functional (and FG) materials fabrication which are free from the abovementioned drawbacks. A hybrid technology is based on the combination of traditional techniques (solid-phase synthesis, hot isostatic pressing (HIP)) and SLS/M technologies and uses dispersion—nanocrystallite powders and CAD of specific (M/NEMS) devices on the base of obtained materials [28, 29].

Fundamentals of the active elements creation were developed at the Samara branch of LPI [25] from functional (smart) materials by the SLS method with a hybrid combination of some traditional processes. The laser influence (LI) on multicomponent (including reaction capable) powdered compositions was studied. Well-known piezoelectric, hexaferrite, and high-temperature superconductivity (HTS) systems ($PbTi_{1-x}Zr_xO_3$—named as PZT; $Li_{0.5}Fe_{2.52x}Cr_xO_4$ and $BaFe_{12-x}Cr_xO_{19}$_spinels; $SrFe_{12}O_{19}$ and $CoFe_2O_4$—HTS) were fabricated [26, 30]. A hybrid layerwise SLS-SHS process (SHS is self-propagated high-temperature synthesis) was realized by means of the laser-controlled combustion reaction inside the oxide stoichiometric mixtures of the above said systems. The main achievement was a production of the 3D parts during the SLS-SHS hybrid process, as well as determination of optimal regimes for their following annealing and polarization (magnetization). The X-ray analysis of the sintered and annealed samples revealed the main phases, responsible for the ferroelectric, antiferroelectric, piezo- (magneto) electric activity of these ceramics. The possibility of association of several approaches (we used PZT as a filler for poly(vinylidene fluoride) (PVDF) polymer) into a united technological process for layerwise syntheses of the FG structures and 3D parts with ferroelectric characteristics of different types of connectivity were also shown [28, 29]. The use of nano-PZT particles in PVDF matrix (which has its own piezoelectric properties) will allow to create such types of connectivity into hybrid AT, which do not exist in the nature, but can possess unique features. Regrettably, density of the synthesized ceramics reached only 3–4 g/cm^3 (that makes up ~40–50% from the theoretical value for PZT) and, as an effect, instead of completely synthesized products, we received only mixture of the initial oxides with partly formed given active phases.

So, the development of the hybrid SLS-HIP technology facilitates manufacturing of ferroelectric ceramics hardly obtainable by conventional methods, of multiferroic and FG materials with giant macroresponses including infinite anisotropy of properties and ultrahigh working temperatures.

5. SLS/M formation of local zones of given configuration possessing magnetic properties

Mechanisms of formation of local zones of a specified configuration with magnetic characteristics during the melting (SLM/S) of nanoparticles additives of transition and/or rare-earth metals with a HTS ceramics in a polymer matrix are of great interest [11, 31].

However this issue was not given a sufficiently thorough study yet. Parts and/or tools of a complex configuration with selective local magnetic properties and multidirectional magnetic poles, which cannot be obtained by the machine treatment, can be easily reproduced by the SLM process. But the mechanisms of formation of such zones and the interaction of materials at the zones boundary have been studied partially only.

If at the moment of a rapid crystallization from the liquid phase into the solid one, the melted pool is influenced with a strong static magnetic field, then the resultant alloy microcrystallines will be arranged along the power lines and after cooling will keep their magnetization. This approach allowing to create local magnetic zones inside a solidified matrix in the longitudinal, transverse, or vertical direction, was proposed by us earlier [32]. By means of the passage by passage consistent cladding, it is possible to form in the vertical direction an overall solid coating of 3D parts with magnetic properties. The obtainment of the 3D part with volume zones of the predetermined magnetic properties is a challenge in materials science and manufacturing. The microstructure of the resultant material is determined by the joint mutual influence of the processes of rapid solidification and crystallization, directional cooling and phase transitions caused by repeated thermal cycles, and chemical composition of the initial powders.

The rapid solidification may cause a volumetric heterogeneity of chemical elements that can lead to the metastable phases formation. The directional heat removal can determine both the preferred direction of the grains growth and crystallographic texture, thereby affecting the magnetic properties of the part [32]. The LI enables to choose the regimes of energy influence on the cladded layers of materials, ensuring the maintenance of the crystalline structure and a given grains size. The orientation of the grains in an external magnetic field characterizes the residual magnetic properties of the obtained material.

Other interesting tasks are the obtainment on nonmagnetic (polymer) substrates of quasi-zero-dimensional points (local zones) [11], quasi-one-dimensional magnetic passages from rare-earth nanoadditives (e.g., Samarium-based alloys) with different composition, two- and three-dimensional arrays based on them; determination of their morphology, crystalline and magnetic structure, the saturation magnetization, residual magnetization, and coercive force.

In the paper [33], a model was considered allowing numerical determination of a resulting field velocity under the laser melting of aluminum in the external magnetic field. The model included heat-dependent characteristics of the material (surface tension and viscosity). A heterogeneous distribution of the magnetic flow density was determined by the experimental Hall data measured for the prototype. It was shown that a constant magnetic flow applied coaxially with the LI, exerts its influence upon the direction of the melted material flow and can be explained as a heterogeneity of electromagnetic destruction.

6. Osteoconductive bioresorbable implants, fabricated via 3D printing with nanoadditives

In the process of fabrication of the source materials for the bone implants capable of providing a sufficiently reliable biological integration with the bone tissue, it is important to search for new phase composition of these materials, to improve their microstructure, to create new architectural types of a macroporous structure, to develop approaches for generating materials with a specified surface roughness of the material, and macropores surface of the material. The material science aspect of the problem is connected with the choice of the chemical and phase composition of the composite based on of the synthesized powders. The backbone of the composite should be a biopolymer degradable in the body environment. The biopolymer can also be with functional nanoadditives [34].

One more method of obtaining strong implants by means of the 3D printing is their hardening by ceramic nanoadditives (Al_2O_3, ZrO_2, AZO, HA) [14, 18, 19, 35–38]. The crucial parameter that determines the suitability of this method for making real bone implants and tissue engineering scaffolds is the print resolution. The polymer matrices framing the macropores must be of a specific architecture that, under the given pores fraction, ensures the following: (a) maximizing of permeability, (b) maximizing of the mechanical characteristics such as strength, hardness (elastic modules), (c) obtaining of the surface where to the cells of osteogenic type could be attached, divided, and differentiate. The task of the powder synthesis of spatially ordered complex structures with a connected system of macropores (not smaller than 100 μm) is quite solvable. These structures determine the osteoconductive properties of the implant.

The best resorption characteristics are observed for tricalcium phosphate $\beta\text{-}Ca_3(PO_4)_2$ (β-TCP, Ca/P = 1.5) [39–43]. The increase of the resorption speed and level, obtaining of a more available environment for the bone tissue development can be achieved by lowering of the Ca/P ratio values below 1.5, while the pH level of the implant environment is maintained close to neutral. A further increase of the resorption limit and speed is associated with the decrease of the Ca/P ratio, i.e., in the transition to the materials, including calcium phosphates phases with condensed phosphate ions, i.e., calcium pyrophosphate and polyphosphate. The presence of condensed phosphates improves the surface hydrophily of the composite and promotes the adsorption of special signaling proteins from the interstitial fluid, resulting in the acceleration of the implant integration in the body.

The task of forming a surface roughness is solved by regulating the system composition, CAD structures and choice of the SLS/M regimes which ensures the fabrication of the specific microporous surface. It is known that smooth matrixes ensure a high proliferative osteoblast potential, while the osteogenous cytodifferentiation is hampered [18]. The increase in the 3D print resolution for matrix in the form of a filled polymer with a given architecture, and then with a given microstructure and phase composition in a porous ceramic material is an important problem [44]. The surface modification by nanoparticles is a perspective approach to the microstructure management for different types of functional implants and tissue engineering scaffolds.

Permeability optimization with the conservation of a sufficient toughness can be realized by the directed obtainment of the given porosity architecture of a 3D part. Topological structures (3D minimum surfaces) occurring in nature ensure the achievement of the maximum permeability under the maximum toughness obtainable for the porous samples. Under the comparable porosity (50%) the most permeable models are cubic, tetrahedral and gyroidal cell models [45, 46]. The gyroidal model has a reasonable compromise between permeability and toughness. Roughness can be introduced as a term with higher angular frequency. This will change the curvature locally, as required for the optimal cell adhesion and growth. A porosity gradient can be easily modeled by adding a linear term. Algebraic form for the function describing the gyroid surface is not complicated and can be represented as trigonometric function, thus allowing the generation and scaling of computer models for such architectures.

Osteoconductivity is the ability of the material to provide the possibility of biological flows, intergrowth of blood vessels into the implant (vascularization), adhesion and binding of osteogenic cells [47]. These characteristics are correlated with the physical permeability of a porous body (bone implants or tissue engineering scaffolds) and are provided by means of the inherent bimodal porosity.

Permeable pores of a large size provide permeability for the flow of necessary biological substances, while the pores of a small size are accountable for the roughness of the surface, giving the signal to spreading and proliferation of bone cells [35, 48, 49]. In order to fabricate a chaotic macroporous structure, different approaches could be used, but only the regular spatial architecture of porous-structured materials allows to increase the permeability and strength of the product to the desired values. The task of designing a regular architecture of the porous space can be only solved with the use of additive technologies, the 3D printing in particular.

Tissue engineering scaffolds possess a certain structured organization, capable to form a framework of life space for the bone-cells predecessors and thus stimulate their functional activity. The scaffolds themselves or in combination with other components exert a direct regulatory effect on the cells predecessors and hereby induce the osteogenesis within the implantation zone. The design of 3D scaffolds must stabilize mechanical loads in the place of contact and prevent the formation of a fibrous capsule around the implant [50].

7. 3D printing of magnetic polymer nanocomposites for medical applications

Magnetic nanoparticles in their pure form are rarely used for therapeutic purposes. Usually, they are encapsulated and/or placed in biologically inert matrixes (oligomers or polymers, including those of natural origin) with the view of reducing a possible toxic influence of the magnetic phase, raising its physicochemical stability and creating of the immobilization conditions on the surface of drugs capsules or matrixes. This problem is successfully solved by the AT methods [51]. The capsulation is usually conducted in ultrafine ferri-, ferro- and superparamagnetic particles, containing stabilizing reagents called "magnetic liquids" [52, 53]. Magnetic nanodots are essentially the same class of nanoparticles, that are respondent to the magnetic field influence. Such particles usually consist of magnetics, such as iron, nickel, and cobalt, and of their chemical mixtures. The increased interest to these objects is explained by their possible application in the catalysis, biomedicine, magnetic-resonance (MR) spectroscopy, and data storage.

Physicochemical characteristics of the magnetic nanoparticles are strongly dependent on the method of their preparation and chemical structure. In most cases, these are particles of the size ranging from 1 up to 100 nm with the apparent superparamagnetism.

During their motion by the blood flow, the nanoparticles could become coated with the blood plasma protein, or absorbed by the immune protectors (macrophages). In order to extent the nanoparticles lifetime in the organism, the polymeric chains are fastened to the nanoparticles. Another method consists in the attachment to nanoparticles of antibodies for malignant tumorous cells since they "know" the pathway to the target (the cancer tumor). The magnetic nanoparticles behavior inside the organism is caused by the surface phenomena, size of nanoparticles and their magnetic characteristics (magnetic moment, remanent magnetism) [54]. The surface phenomena chemistry is particularly important for the elimination of the influence of reticular-endothelial system (RES) being the part of the immune system, and for the elongation of the nanoparticles lifetime within the bloodstream. The nanoparticle coating by a neutral and hydrophilic compound (i.e., polyethylene glycol (PEG), polysaccharides, etc.) enlarges the circulating time of the particle existence from minutes to hours and even days [55].

One more possible way is the reduction of the particles size. However in spite of all the efforts, the RES effect has not been completely avoided, and toxicological problems due to the undesirable displacement into other organism areas are still remaining.

Magnetic nanoparticles have found a lot of efficient applications in biomedicine that open new possibilities for therapy and diagnostics of a number of heavy diseases. Alive organisms are built of cells with a typical size of about 10 μm. And in turn, the cell components are much smaller and have the size less than 1 μm. For medical applications, it is important that nanoparticles have controllable sizes within the range of several nanometers to hundreds nanometers, which are comparable with the sizes of intracellular biological objects—(10–100 nm), viruses (20–450 nm), proteins (5–50 nm), and genes (about 2 nm in the transverse direction and 10–100 nm lengthwise). By their size (from 4 to 1000 nm) and their mass, nanoparticles are intermediate between molecules and alive cells.

This size range opens great freedom for promissory medical application of nanoparticles at the AT.

7.1. Magnetic hyperthermia

Magnetic nanoparticles can be incorporated into a bioresorbable polymer matrix so that they could reverberatory respond on the alternating external magnetic field (EMF) of a certain frequency and amplitude, and concurrently effectively absorb the EM energy and transfer it as a heating to the surrounding biological tissues. For instance, a magnetic nanoparticle can be used as hyperthermia agent heated in the applied EMF and delivering mortal doses of thermal energy to the tumors cells [56]; or as facility capable of increasing the efficiency of chemotherapy, beam and laser therapy, where magnetic nanoparticles lead to a moderate degree of heating resulting in a more efficient destruction of the malignant tissue. For the magnetic hyperthermia the particles are to possess high SAR (specific absorption rate) allowing their fast heating in the alternating magnetic field. The cancerous cells are known to be ruined under 42–43°C.

Study on the absorption rate of the EME by the magnetic liquid SAR with nanoparticles is important for the certification of fluid-magnetic hyperthermia (FMH) drugs. Along with the parameters, such as Curie point, value of saturation-specific magnetization, and toxicity, the SAR as an attribute of the magnetic liquid determines the possibility and efficiency of medical applications. For the FMH, it is necessary to develop in polymers the magnetic nanoparticle compositions, capable of releasing a dosated portion of magnetic particles. At present, superparamagnetic particles on the magnetite base have been developed, with the substitution of Fe^{2-} by manganese and zinc, and Fe^{3-}—by the gadolinium. At a slight decrease in the specific magnetic receptivity and specific magnetization saturation, the Curie temperature reduction from 575°C (magnetite) to 70°C was achieved. A similar problem was also solved for manganites.

The tendency to reducing the size resulted in a dominating use of superparamagnetic systems (for instance, superparamagnetic iron nano-oxides (SPIO)) without a hysteresis loop. The magnetic nanoparticles used for therapeutic purposes, can consist of ferromagnetic, ferrimagnetic, or superparamagnetic materials. Their main advantage is the possibility of contactless control of their displacement in the organism under the EMF. In our researches, it was shown that promissory materials are iron oxide nanoparticles with spinel structure (magnetite, maghemite) and even high-temperature superconducting ceramics ($SrFe_{12}O_{19}$) [31].

7.2. Radio-absorbing coatings from polymers with nano inclusions

To protect against the EMF, radioabsorbed materials (RAM) are widely used, in particular polymer composites with ferrimagnetic or ferromagnetic nanoadditives are used as RAM. Efficient RAM should satisfy a number of requirements: maximum absorption of electromagnetic waves within a wide frequency range, minimum reflection and lack of harmful fumes, fire safety, small dimensions, and light weight.

Two types of RAM are distinguished: interference type, wherein the electromagnetic waves are weakened due to the destructive interference, and absorbing type in which the electromagnetic energy is converted into thermal one due to scattered currents, dielectric or magnetic relaxation. Depending on the main source of relaxation, the RAM can be of dielectric or magnetodielectric types, and depending on the operating range of the absorption frequencies, there exist narrow band or wide band RAM.

As the alternative, the transparent in visual light materials (glass, polymer) can be used with embedded magnetic nanoparticles (particularly, magnetite Fe_3O_4 nanoparticles and hexagonal strontium ferrite $SrFe_{12}O_{19}$ nanoparticles) [11, 31]. It is known that spinel ferrites and their composites effectively absorb EMR within the frequency band of 0.8–2.0 GHz. Hexagonal ferrites intensively absorb in the range of tens of GHz. So, by combining the nanoparticles of magnetite and hexaferrites (e.g., strontium hexaferrite), it is possible to improve the RAM performance at the edges of the frequency band, i.e., at few GHz and at tens GHz, thus significantly increasing the operating range of RAM.

Earlier, we have obtained the 3D parts from polymers with nano inclusions of Fe_3O_4 and $SrFe_{12}O_{19}$ by the SLS/M. By the proposed method, it is possible to fulfill easy polymer coatings with nanoparticles of the above-mentioned ferrites on the surface of medical equipment thus providing its protective covering [57, 58].

7.3. Magnetic nanoparticles for targeted drug delivery systems

The targeted drug delivering (TDD) is one more promissory sphere for the medical application of magnetic nanoparticles [49]. Its main advantages are significant reducing of the drug toxic effect on other organs and systems, possibility to direct and retain the drug-containing nanoparticles in a certain place with the help of magnetic field, and to visualize them by the magnetic resonance imaging (MRI) methods. An important property of magnetic nanoparticles is a possibility of their local heating by the high-frequency magnetic field for the initiating of the drug desorption/decapsulation mechanism or for magnetic hyperthermia. Usually for the TDD, the superparamagnetic particles are used as the magnetic carrier since after the magnetic field influence they are not aggregating. However, there exists the problem of the reduction of magnetic influence power after the drug delivery, resulting in complicating of the particles retaining in close proximity from the target object, particularly under the powerful blood flow effect.

For a long time, already various one- and multicomponent liposomes generating in lipid solutions have been well known. For the practical purposes, liposomes with sizes of 20–50 nm are of interest since they are used as drug delivering systems to a biological target [59]. There is also a whole set of natural nanocarriers, for instance, viruses. Especially, processed adenoviruses can effectively be used for vaccination through the skin. Other examples of biogenic nanoparticles capable of the targeted delivering are lipid nanotubes, nanoparticles and nanoemulsion of natural origin, some cyclical peptides, chitozans, and nucleic acids [60].

"Magnetic" bacteria can deliver drugs, for instance they can be used as a system for the point delivery of drugs to the diseased tissues also. The MC-1 bacteria are capable of a fast moving owing to the rotating of their own flagellums. Besides, they contain magnetic nanoparticles and are therefore sensitive to the magnetic field and are enforced to move along the lines of force. As a generator of these lines of force the MRT device, for instance, can serve.

Nanospheres and nanocapsules are referred to the polymer nano objects. The nanospheres are solid matrixes with an active material spread over their polymer surfaces, while in nanocapsules the polymer coatings form the cavity filled with liquid. Hence, the active material is released into the organism by different mechanisms—the release from the nanospheres is of an exponential nature, whereas from the nanocapsules it happens with a constant velocity and for a long time. Polymer nanoparticles could be received via the SLS/M process as core/shell composites. They are polysaccharides, polyglycolic acids, polylactides, polyacrylates, acrylic polymers, polyethyleneglycol (PEG), etc. Polymer materials are characterized by a whole number of useful properties for the medical transportations, such as biocompatibility, ability to be biologically decomposed, and multifunctionality.

Dendrimers are becoming of a high interest. This is a new type of polymers, having a branching, dendrite-like structure instead of the usual linear one. Dendrimers are often mentioned in the context of their nanotechnological and medical applications. Dendrimers are a unique type of polymers since their size and shape can be superaccurately specified under the chemical synthesis that is extremely important for the TDD. Dendrimers are obtained from monomers by conducting their consequent convergent and divergent polymerization (including the use of the peptide synthesis methods), thus determining the pattern of branching.

Typical monomers used in the synthesis are polyamide-amine and amino acid—lysine. The "target" molecules are linked with dendrimers either by means of forming complexes with their surface, or by means of embedding deep between their separate chains. It is also possible to place in a stereo-specific way the necessary functional groups on the dendrimers surface so that they could interact with viruses and cells.

Carbon nanoparticles (fullerenes and nanotubes) are well known material for the TDD. Nanotubes can be used as microscopic containers introduced by the SLS/M process for the transporting of many chemical or biologically active materials: proteins, poisonous gases, fuel components, and even melted metals. For medical applications, the nanotubes possess an important increased affinity to lipid structures, they are able to form stable complexes with peptides and DNA-oligonucleotides and even encapsulate these molecules. Taken together, all these characteristics and properties provide their application as efficient drug delivering systems for vaccines and genetic material.

Promissory platform technologies are microcapsulation, technology for manufacturing of matrix, multilayered, and coated tablets and capsules. Designed and described in the literature are the platform technologies for making the nanosize complexes of active materials with biocompatible and biodegradable synthetic and natural polymers [6, 18]. The nanosize may result in the increase in a drug activity by several times and also in the reinforcement of therapeutic characteristics. Preclinical studies of known drugs in new nanopacking are already carried out (e.g., taksol or nurophene of a prolongation effect). Platform technologies for a controllable drugs release are very important for the targeted delivery of high toxic antineoplastic drugs.

Traditional cancer drugs are evenly distributed by the whole organism, thus reaching both the diseased and healthy organs. This problem can be solved with the help of directed DDS (drug delivery system) when the drug is delivered along with a biodegradable polymer used as transport. In this case, the drug is released not immediately, but gradually, in the process of the polymer degradation.

There exist the target drugs delivering by nanoparticles of a genetic material, DNA or RNA. For the magnetic target influence, the drug or therapeutic radionuclide is attached to magnetic compound, which is entered into the organism, and then concentrated in the area of the target influence by means of a magnetic field (here the implanted constant magnet or the EMF could be used).

The magnetic DDS in its current state is mostly applicable to well-studied tumors, whereas the medical treatment of metastatic and/or small tumors at their early stage of development still remains an unsolved problem. The therapy of emerging tumors implies the development of a new generation of nanoparticles of teradiagnostics type, capable of recognizing small clusters of cancerous cells and delivering the elements (drugs or hyperthermia materials) required for their destruction. Teradiagnostics will ensure a permanent control of the course of treatment with a concurrent checkup of the drug delivery results and influences.

7.4. Magnetic resonance imaging

Lately, a number of directions in the molecular visualization sphere have got a wide development via the 3D printing. MRI is one of the most powerful noninvasive methods of visualizations intensively used in clinical medicine [61]. Contrast agents based on the magnetic nanoparticles are one of the most perspective medical applications and are considered as the following MRI-agents generation. Magnetic SPIO are also actively researched as contrast agents for the MRI. New biological applications of these contrast agents are the following: checking of a bloodstream, cancer and tissue engineering scaffolds agents, the MRI contrasts, checking the cells motion, and biomolecular study. In contrast to paramagnetic ions, the SPIO nanoparticles have a higher molecular relaxation. Therefore, in case of their use under low concentration for the bloodstream checkup, they can have some advantages. The MRI yields to positron spectroscopy in its resolution but it gains in its cost. An inexpensive MRI procedure with the SPIO nanoparticles use, distinguishes tumor metastasis of approximately 1 mm size.

8. Conclusions

In this chapter, we reviewed the modern tendency to fabricate the 3D printing of functionally graded structures from the polymer-covered nanocomposites: inkjet 3D printing of optical nanostructures; 3D-printing of polymer nanocomposites with specific viscosity, osteoconductivity and biodegradability; FG nanostructures with different connectivity types, obtained by a hybrid SLS technology; opportunities of the SLS/M formation of magnetic local zones of the specified configuration; and 3D printing of magnet polymer nanocomposites/objects for medical applications.

Numerous problems have been identified and some solutions have been proposed and assessed. Still, a lot of further researches are required to implement a true industrial process, providing the rapid fabrication of high quality net-shaped MEMS/NEMS devices—implants, catalytic filters and membranes, lab-on-chips and drug delivery systems, magnetic hyperthermia systems and 3D crystalline structures for hydrogen storage, and/or optical devices.

Acknowledgements

Prof. I. Shishkovsky is grateful for the financial support provided by the Russian Science Foundation (grant no. 15-19-00208).

Author details

Igor Volyanskii[1*] and Igor V. Shishkovsky[1,2*]

*Address all correspondence to: i.volyansky@gmail.com and shiv@fian.smr.ru

1 Samara State University, Samara, Russian Federation

2 Lebedev Physical Institute (LPI) of Russian Academy of Sciences, Samara Branch, Samara, Russian Federation

References

[1] Vaezi, M., Yang, S. Freeform fabrication of nanobiomaterials using 3D printing. Rapid Prototyping of Biomaterials: Principles and Applications, Edited by: R. Narayan, Woodhead Publishing Limited, Cambridge, UK 2014, p. 16–74.

[2] Carrow, J.K., Kerativitayanan, P., Jaiswal, M.K., Lokhande, G., Gaharwar, A.K. Polymers for bioprinting. Essentials of 3D Biofabrication and Translation, Edited by Anthony Atala and James J. Yoo, Elsevier Inc., Oxford, UK, 2015, p. 229–248.

[3] Shishkovsky, I.V., and Morozov, Yu.G. Multilayer polymer structures containing Ni/Cu nanoclusters as prepared by selective laser sintering. International Journal of Self Propagating High Temperature Synthesis. 2011, Vol. 20, No. 1, p. 53–60.

[4] Kim, G.H., Son, J.G. 3D polycarprolactone (PCL) scaffold with hierarchical structure fabricated by a piezoelectric transducer (PZT)-assisted bioplotter. Applied Physics A: Materials Science and Processing, Vol. 94, No. 4, 2009, p. 781–785.

[5] Shishkovsky I.V., Morozov Yu.G. Electrical and magnetic properties of multilayer polymer structures with nano inclusions as prepared by selective laser sintering. Journal of Nanoscience and Nanotechnology, Vol. 13, No. 2, 2013, p. 1440–1443

[6] Shishkovsky I.V., Morozov. Y.G. Laser assisted fabrication of porous polymer MEMS with nano structured additives. MRS Proceedings, 2011, 1312, p. 343–348.

[7] Shishkovsky I.V., Bulanova A.V., Morozov, Y.G. Porous polycarbonate membranes with Ni and Cu nano catalytic additives fabricated by selective laser sintering. Journal of Materials Science and Engineering B, Vol. 2, No. 12, 2012, p. 634–639.

[8] Gibson, I., Rosen, D.W., Stucker, B. Additive Manufacturing Technologies. Rapid Prototyping to Direct Digital Manufacturing. Springer, Heidelberg, Germany, 2010.

[9] Shishkovsky, I., Scherbakof, V., Volyansky, I. Low-dose laser sintering of Cu nanoparticles on the ceramic substrate during ink-jet interconnection. Proceedings of SPIE, Vol. 9065, Fundamentals of Laser-Assisted Micro- and Nanotechnologies, 2013, 90650I.

[10] Billiet, T., Vandenhaute, M., Schelfhout, J., Van Vlierberghe, S., Dubruel, P. A review of trends and limitations in hydrogel-rapid prototyping for tissue engineering (review). Biomaterials, Vol. 33, No. 26, 2012, p. 6020–6041.

[11] Shishkovsky I.V., Scherbakov V.I., Morozov Yu.G. Nanocomposites of polyetherketone/Fe_xO_y nano oxides processed by selective laser sintering (poster). XII International Conference on Nanostructured Materials (NANO 2014). July 13–18, 2014, Moscow, Russia, p. 805.

[12] Chung, J.H.Y., Naficy, S., Yue, Z., Kapsa, R., Quigley, A., Moulton, S.E., Wallace, G.G. Bio-ink properties and printability for extrusion printing living cells. Biomaterials Science, Vol. 1, No. 7, 2013, p. 763–773.

[13] Shishkovsky I.V, Juravleva I.N. Kinetics of polycarbonate distraction during laser-assisted sintering. International Journal of Advanced Manufacturing Technology, Vol. 72, 2014, p. 193–199.

[14] Shishkovsky I., Nagulin K., Sherbakov V. Study of biocompatible nano oxide ceramics, interstitial in polymer matrix during laser-assisted sintering. International Journal of Advanced Manufacturing Technolog, Vol. 78, No. 1–4, 2015, p. 449–455.

[15] Lebel, L.L. and Therriault, D. Multiscale Manufacturing of Three-Dimensional Polymer-Based Nanocomposite Structures, In: Reddy, B. (Ed.), Advances in Diverse Industrial Applications of Nanocomposites, InTech Publisher, Rijeka, Croatia, 2011.

[16] Bose, S., Vahabzadeh, S., Bandyopadhyay, A. Bone tissue engineering using 3D printing. Material Today, Vol. 16, No. 12, 2013, p. 496–504.

[17] Carrico, J.D., Traeden, N.W., Aureli, M., Leang, K.K. Fused Filament Additive Manufacturing of Ionic Polymer–Metal Composite Soft Active 3D Structures. ASME 2015 Conference on Smart Materials, Adaptive Structures and Intelligent Systems, Colorado Springs, Colorado, USA, September 21–23, 2015, Paper No. SMASIS2015-8895, p. V001T01A004.

[18] Shishkovsky, I.V., Volchkov, S.E. Ceramics-filled 3D porous biopolymer matrices for tissue-engineering on the stem cell culture: benchmark testing. In: Bartolo et al. (Eds.), High Value Manufacturing: Advanced Research in Virtual and Rapid Prototyping, 2014, Taylor & Taylor & Francis Group, London, UK, p. 121–126.

[19] Shishkovsky, I., Volchkov, S. Influence of the laser assisted fabricated 3D porous scaffolds from bioceramoplasts of micron and nano sizes on culture of MMSC. Proceedings of SPIE, Vol. 9065, Fundamentals of Laser-Assisted Micro- and Nanotechnologies, 2013, p. 906515.

[20] De Santis, R., Gloria, A., Russo, T., D'Amora, U., Zeppetelli, S., Dionigi, C., Sytcheva, A., Herrmannsdörfer, T., Dediu, V., Ambrosio, L. A basic approach toward the development of nanocomposite magnetic scaffolds for advanced bone tissue engineering. Journal of Applied Polymer Science, Vol. 122, No. 6, 2011, p. 3599–3605.

[21] Ivanova, A.M., Kotova, S.P., Kupriyanov, N.L., Petrov, A.L., Tarasova, E.Yu, Shishkovskii, I.V. Physical characteristics of selective laser sintering of metal–polymer powder composites. Quantum Electronics, Vol. 28, No 5, 1998, p. 420–425.

[22] I.V. Shishkovskii, Kupriyanov, N.L. Thermal fields in metal–polymer powder compositions during laser treatment. High Temperature, Vol. 35, No. 5, 1997, p. 710.

[23] Shishkovsky, I.V. Thermoviscoplasticity of powder composition under selective laser sintering. In: , Panchenko, V.Y., Golubev, V.S. (Eds.), Proceedings of SPIE, Vol. 4644-71, April 2002, p. 446–449, Seventh International Conference on Laser and Laser-Information Technologies.

[24] Shishkovsky, I.V. Rheological dynamics of powder compositions during selective laser sintering process. In: Geiger, M., Otto, A. (Eds.), Proceedings of the 3rd International Conference Laser Assisted Net Shape Engineering LANE'2001, 28–31 August 2001, Erlangen, Germany, p. 399-406.

[25] Shishkovsky, I.V. Laser Synthesis of Functional Mesostructures and 3D Parts. Fizmatlit Publ., Moscow, 2009, ISBN 978-5-9221-1122-5, 424 p.

[26] Gureev, D.M., Ruzhechko, R.V., Shishkovskii, I.V. Selective laser sintering of PZT ceramic powders. Technical Physics Letters. Vol. 26, No 3, 2000, p. 262–264.

[27] Tarasova, E.Y., Kryukova, G.V., Petrov, A.L., Shyshkovsky, I.V. Structure and properties of porous PZT ceramics synthesized by selective laser sintering method. In: Helvajian, V.H., Sugioka, K., Gower, M.C., Dubowski, J.J. (Eds.), Proceedings of SPIE.

Vol. 3933-64. June 2000, p. 502–504, Laser Applications in Microelectronic and Optoelectronic Manufacturing.

[28] Shishkovsky, I.V., Kuznetsov, M.V., Morozov, Yu.G. Layering fabrication, structure and electromagnetic properties of perovskite phases by hybrid process: self-propagated high-temperature synthesis and selective laser sintering. Phase Transitions – A Multinational Journal, 2013, Vol. 86, No. 11, p. 1085–1093.

[29] Tarasova, E., Juravleva, I., Shishkovsky, I., Ruzhechko, R. Layering laser-assisted sintering of functional graded porous PZT ceramoplasts. Phase Transitions – A Multinational Journal, Vol. 86, No. 11, 2003, p. 1121–1129.

[30] Shishkovskii, I.V., Kuznetsov, M.V., Morozov, Yu.G. New methods for development of three-dimensional ceramics based on barium hexaferrite with chromium additives. Glass and Ceramics, Vol. 60, No. 5–6, 2003, p. 174–178.

[31] Shishkovsky, I.V., Scherbakov, V.I., Kuznetsov, M.V. Study of core–shell SHTC/polycarbonate covered with ultrafine particles fabricated by laser assisted sintering (poster). XII International Conference on Nanostructured Materials (NANO 2014). July 13–18, 2014, Moscow, Russia, p. 986.

[32] Saphronov, V., Shishkovsky, I. Peculiarities of selective laser melting process for permalloy powder. Materials Letters, Vol. 171, 2016, p. 208–214.

[33] Gatzen, M., Tang, Z. CFD-based model for melt flow in laser beam welding of aluminium with coaxial magnetic field. Physics Procedia. Vol. 5, 2010, p. 317–326.

[34] Seliktar, D., Dikovsky, D, Napadensky, E. Bioprinting and tissue engineering: recent advances and future perspectives (review). Israel Journal of Chemistry , Vol. 53, No. 9–10, 2013, p. 795–804.

[35] Shishkovsky, I.V., Pitrov, V.S., Kuznetsov, M., Morozov, Yu., Volova, L., Barikov, I., Fakeev, S. Porous surface structure of biocompatible implants and tissue scaffolds base of titanium and nitinol synthesized SLS/M methods. In: Panchenko, V., Larichev, A., Zheltov, G. (Eds.), Proceedings of SPIE – Vol. 6734-22, International Conference on Lasers, Applications, and Technologies 2007: Laser Technologies for Medicine, 67340N, August 1, 2007.

[36] Zheng, H., Zhang, J., Lu, S., Wang, G., Xu, Z. Effect of core–shell composite particles on the sintering behavior and properties of nano-Al_2O_3/polystyrene composite prepared by SLS. Materials Letters, Vol. 60, No. 9–10, 2006, p. 1219–1223.

[37] Zhang, J., Zheng, H.-Z., Xu, Z.-F., Sun, S.-W., Liu, Y.-H. Study on characterization of core-shell nano-Al_2O_3/PS composite particles and toughening polystyrene prepared by SLS. Journal of Materials Engineering, No. 3, 2007, p. 24–27.

[38] Shishkovsky, I., Scherbakov, V. Selective laser sintering of biopolymers with micro and nano ceramic additives for medicine. Physics Procedia, No. 39, 2012, p. 491–499.

[39] Shishkovsky, I.V., Tarasova, E.Yu., Zhuravel', L.V., Petrov, A.L.. The synthesis of a biocomposite based on nickel titanium and hydroxyapatite under selective laser sintering conditions. Technical Physics Letters, Vol. 27, No. 3, 2001, p. 211–213.

[40] Tan, K.H., Chua, C.K., Leong, K.F., Cheah, C.M., Cheang, P., Abu Bakar, M.S., Cha, S.W. Scaffold development using selective laser sintering of polyetheretherketone–hydroxyapatite biocomposite blends. Biomaterials, Vol. 24, No. 18, August 2003, p. 3115–3123.

[41] Eosoly, S., Lohfeld, S., Brabazon, D. Effect of hydroxyapatite on biodegradable scaffolds fabricated by SLS. Key Engineering Materials, Vol. 396–398, 2009, p. 659–662.

[42] Castilho, M., Moseke, C., Ewald, A., Gbureck, U., Groll, J., Pires, I., Teßmar, J., Vorndran, E. Direct 3D powder printing of biphasic calcium phosphate scaffolds for substitution of complex bone defects. Biofabrication, Vol. 6, No. 1, 2014, Article number 015006.

[43] Duan, B., Wang, M., Zhou, W.Y., Cheung, W.L., Li, Z.Y., Lu, W.W. Three-dimensional nanocomposite scaffolds fabricated via selective laser sintering for bone tissue engineering. Acta Biomaterialia, Vol. 6, No. 12, 2010, p. 4495–4505.

[44] Hollister, S.J. Porous scaffold design for tissue engineering. Nature Materials, Vol. 4, No. 7, 2005, p. 518–524.

[45] Shishkovsky, I.V. Correlation of laser design-microstructure-properties in porous 3D matrix for tissue engineering and drug delivery systems. Issues Samara Science Center of RAS, Vol. 13, No. 4, 2011, p. 45–53.

[46] Dias, M.R., Fernandes, P.R., Guedes, J.M., Hollister, S.J. Permeability analysis of scaffolds for bone tissue engineering. Journal of Biomechanics, Vol. 45, No. 6, 5 2012, p. 938–944.

[47] Heo, S.-J., Kim, S.-E., Wei, J., Hyun, Y.-T., Yun, H.-S., Kim, D.-H., Shin, J.W., Shin, J.-W. Fabrication and characterization of novel nano- and micro-HA/PCL composite scaffolds using a modified rapid prototyping process. Journal of Biomedical Materials Research – Part A, Vol. 89, No. 1, 2009, p. 108–116.

[48] Shishkovsky, I., Volchkov, S. Influence of the SLS-technique-obtained 3D porous bioresorbable matrix for tissue engineering on culture of multipotent mesenchymal stem cells. 2012 MRS Fall Meeting & Exhibit, November 25–30, 2012, Boston, MA, USA, Symposium L: Biomimetic Nanoscale Platforms, Particles, and Scaffolds for Biomedical Applications, L6.13.

[49] Shishkovsky, I.V. SLS design of porous drug delivery system from porous nitinol. Nano- i mikrosistemnaya tekhnika, Vol. 9, 2012, p. 39–43.

[50] Shishkovsky, I.V., Kuznetsov, M.V., Morozov, Yu.G. Computer-controlled synthesis of orthopedic implants. International Journal of Self-Propagating High-Temperature Synthesis, Vol. 18, No. 2, 2009, p. 137–138.

[51] Shishkovsky, I., Sherbakov, V., Morozov, Yu. Layerwise laser-assisted sintering and some properties of iron oxide core/PEEK shell magnetic nanocomposites. Microelectronic Engineering, Vol. 146, 2015, p. 85–91.

[52] Tóth, B.G., Péter, L., Dégi, J., Bakonyi, I. Magnetoresistance and surface roughness study of electrodeposited $Ni_{50}Co_{50}$/Cu multilayers. Journal of the Electrochemical Society, Vol. 160, No. 8, 2013, p. D307–D314.

[53] Hu, S.-H., Liu, T.-Y., Tsai, C.-H., Chen, S.-Y. Preparation and characterization of magnetic ferroscaffolds for tissue engineering. Journal of Magnetism and Magnetic Materials, Vol. 310, No. 2 Suppl. Part 3, 2007, p. 2871–2873.

[54] Lopez-Lopez, M.T., Scionti, G., Oliveira, A.C., Duran, J.D.G., Campos, A., Alaminos, M., Rodriguez, I.A. Generation and characterization of novel magnetic field-responsive biomaterials. PLoS One, Vol. 10, No. 7, 2015, Article number e0133878.

[55] De Santis, R., Gloria, A., Russo, T., D'Amora, U., Zeppetelli, S., Dionigi, C., Sytcheva, A., Herrmannsdörfer, T., Dediu, V., Ambrosio, L. A basic approach toward the development of nanocomposite magnetic scaffolds for advanced bone tissue engineering. Journal of Applied Polymer Science, Vol. 122, No. 6, 2011, p. 3599–3605.

[56] Shishkovsky, I., Scherbakov, V., Morozov, Yu. Selective laser sintering of PEEK core shell/iron oxide magnetic nanoparticles (oral). Abstracts Book of 9th International Conference on Surfaces, Coatings and Nano-Structured Materials (NANOSMAT 2014), 8–11 September 2014, Dublin, Ireland, p. 283–284.

[57] Shishkovsky, I.V. Chemical and physical vapor deposition methods for nanocoatings. In: Hamdy, A.S., Tiginyanu, I. (Eds.), Nanocoatings and Ultra Thin-Films, 2011, 414 p., Woodhead Publishing Limited, Abington Cambridge, UK, on-line ISBN 978-1-84569-812-6, p. 57–77.

[58] Bhowmick, A., Saha, A., Pramanik, N., Banerjee, S., Das, M., Kundu, P.P. Novel magnetic antimicrobial nanocomposites for bone tissue engineering applications. RSC Advances, Vol. 5, No. 32, 2015, p. 25437–25445.

[59] Sussman, E.M., Jayagopal, A., Haselton, F.R., Shastri, V.P. Engineering of solid lipid nanoparticles for biomedical applications. ACS Symposium Series, Vol. 992, 2008, p. 139–152.

[60] Mazzarino, L., Otsuka, I., Halila, S., Bubniak, L.D.S., Mazzucco, S., Santos-Silva, M.C., Lemos-Senna, E., Borsali, R. Xyloglucan-block-poly(ε-caprolactone) copolymer nanoparticles coated with chitosan as biocompatible mucoadhesive drug delivery system. Macromolecular Bioscience, Vol. 14, No. 5, 2014, p. 709–719.

[61] Krishnan, S.P., Dawood, A., Richards, R., Henckel, J., Hart, A.J. A review of rapid prototyped surgical guides for patient-specific total knee replacement. Journal of Bone and Joint Surgery – Series B, Vol. 94 B, No. 11, 2012, p. 1457–1461.